MOTORCYCLE

EVOLUTION • DESIGN • PASSION

MICK WALKER

MOTORCYCLE
EVOLUTION • DESIGN • PASSION

JOHNS HOPKINS UNIVERSITY PRESS
BALTIMORE

To Gary. A much-missed son and friend

MOTORCYCLE

MICK WALKER

Printed in China on acid-free paper

First published in Great Britain in 2006 by Mitchell Beazley, an imprint of Octopus Publishing Group Ltd., 2–4 Heron Quays, London E14 4JP

North American edition published 2006 by the Johns Hopkins University Press, by arrangement with Mitchell Beazley

9 8 7 6 5 4 3 2 1

The Johns Hopkins University Press
2715 North Charles Street
Baltimore, Maryland 21218-4363
www.press.jhu.edu

ISBN-13: 978-0-8018-8530-3
ISBN-10: 0-8018-8530-2 (hardcover: alk. paper)

Library of Congress Control Number: 2006924782

A catalog record for this book is available from the British Library.

Set in Eurostile

CONTENTS

FOREWORD

I was delighted when Mick Walker invited me to write the foreword for his latest book, *Motorcycle*. It is a comprehensive volume that covers all the notable motorcycles and their designers, from Gottlieb Daimler's crude "bone-shaker" of 1885 up to the sophisticated, high-performance machines of today.

Having been a retailer and agent, racer and builder/developer of specialized racing motorcycles myself, I found it fascinating to read how the skills and ideas of the early designers – such as Alfred Angus Scott, the Collier brothers, Edoardo Bianchi, James Lansdowne Norton, Adalberto Garelli, Carlo Guzzi, Edward Turner, Rex McCandless, and many others – have been built upon by later generations in their constant quest for improvement.

I must admit to being truly amazed, when studying this book, by the sheer breadth and diversity of the subject. While the motorcycles and their designers are central, of course, many giant steps in technical innovation and development have been taken in little over 120 years since Daimler's first two-wheeled prototype appeared.

The early designers were real "artists in metal". Although CAD (computer-aided design) has become a vital tool in the modern motorcycle industry, their ideas, like mine, began with pen and paper or chalk on the floor, followed by hours in the machine shop.

The developments and designs of all the great marques are here, from AJS to Zündapp.

There have been plenty of technical innovations that have played important roles in shaping motorcycle design over the decades. This book examines the most outstanding, such as multi-speed, foot-change gearboxes, methods of lubrication, liquid cooling, final drive, rotary engines, multi-cylinder engines, and turbocharging.

Nowadays, the motorcycle has moved on from its original role of economical personal transport, occupying instead a sophisticated niche in the market of leisure with style. Mick has reflected this shift in the final chapter, "Lifestyle and Leisure", in which he examines custom bikes and retro machines, as well as recent Ducati, Harley-Davidson, and Triumph models, the Yamaha R1, the MV Agusta F4, off-road sport, and the phenomenon of colour coding.

Motorcycle is a thorough and comprehensive book, and it's my pleasure to recommend it.

Colin Seeley

INTRODUCTION

The history of motorcycle design is a fascinating subject, concerning not only the machines themselves, but also the men who created them. Although the motorcycle is as old as its four-wheeled counterpart, far less has been written about its design and development, from the crude motorized two-wheelers of the late 19th century to the 290km/h (180mph) projectiles of the early 21st century. Within the pages of *Motorcycle*, I have attempted to provide a complete insight to this riveting story.

While the German engineer Gottlieb Daimler is credited with producing the first motorcycle powered by an internal-combustion engine, in 1885, there were others who had carried out important work prior to that historic event.

Bicycles came before petrol engines, and steam engines came before both. In fact, the first powered two-wheeler was a cycle equipped with a small steam engine.

The first attempts to provide a two-wheeler with motive power that relied on something other than cranks and pedals occurred during the late 1860s, being inspired by the development of steam power. Steam engines, as epitomized by contemporary rail and road locomotives, were truly massive creations, however, with huge cast-iron cylinders, vast flywheels and heavy connecting rods. Even so, both

Frenchman Pierre Michaux and American Sylvester Howard Roper created steam-driven, two-wheeled machines in 1869. Michaux's device featured a single brazed-steel cylinder, a light steel piston and twin flywheels. Steam was generated by a transverse, multi-tube, cylindrical boiler, heated by alcohol fuel through a series of burners. Roper's machine had a charcoal-fired, twin-cylinder engine. In truth, however, steam power was much more suited to four wheels, because the machinery and boiler could be installed more easily.

During those early pioneering days, inventors experimented with a range of possible power sources for two-wheeled machines. These included compressed air, clockwork, carbonic-acid gas and even hydrogen gas.

The basic principle of the internal-combustion (petrol) engine was outlined by Frenchman Alphonse Beau de Rochas in 1862. Its working application was not patented until 1876, however, by two Germans, Nicolaus Otto and Eugen Langen. The Otto four-stroke cycle of induction, compression, firing, and exhaust remains unchanged today.

Another early pioneer of the motorcycle was Englishman Edward Butler, who employed a two-stroke engine to power a three-wheel machine. This vehicle was patented provisionally in 1884, but it was not actually completed for another four years.

Thus, Gottlieb Daimler and his assistant, Wilhelm Maybach, were responsible for the first real motorcycle. It represented the first step in an incredible journey of technical development.

Daimler also devised a primitive, two-speed transmission. He replaced the existing belt final drive with an internally toothed, rear wheel rim, while a belt primary drive turned a countershaft and pinion that engaged with the toothed rim. To change speed, however, the rider had to stop each time.

Another vital early development was John Boyd Dunlop's patenting of the pneumatic tyre (originally conceived by R.W. Thomson in 1845). Dunlop's son, Johnny, had not been happy with the ride of his little

bicycle, so his father had designed inflatable tyres for it, leading to a new technology that would benefit all road vehicles. Previously, the steel or wooden wheel rims had been in direct contact with the road surface, hence the term "bone-shaker".

The next notable step forward was taken by the German Hildebrand & Wolfmüller company, which was the first to begin series production of a motorcycle – a four-stroke, twin-cylinder model – during the mid-1890s. Although the venture failed, others soon followed the example.

An important advance in ignition was made by Robert Bosch, just before the end of the 19th century, when he developed the low-tension

facing page During the 1950s, the British Panther Model 100 600cc single was a popular choice for sidecar work.

above and right Sales brochures often reflected the motorcycle's sporting achievements or lifestyle contribution. These BSA publications show 1964/65 World Motocross Champion Jeff Smith and a 1968 A65 650cc unit twin.

magneto. At about the same time, Ludwig Rubb and Christian Haab pioneered the use of the engine as a stressed member of the frame.

Single- and twin-cylinder engines were the most common motorcycle power units during the pioneer period, but a British Army officer, Major Henry Capel Lofft Holden, was responsible for the world's first four-cylinder machine, in 1896. His engine featured two sets of twin cylinders and was mounted low in the frame. A water-cooled version arrived in 1899.

In the main, the early years of the motorcycle were dominated by German, French, and Belgian designs. The Americans and British made up for lost time in the first decade of the 20th century,

however, when dozens of manufacturers, large and small, entered the motorcycle market.

At first, single cylinders, single speeds, and belt drives were the norm, but there were enough innovative engineers to stimulate interest in advances. These included overhead valves, clutches, and variable gears, unit construction, shaft drive, and sprung frames. In fact, the vast majority of modern motorcycle systems were conceived many years ago; only the lack of suitable materials, high costs, and natural resistance to the unproven and unorthodox often slowed their introduction.

American pioneers included Albert Pope, Hiram Maxim, George M. Holley, and a Swedish immigrant,

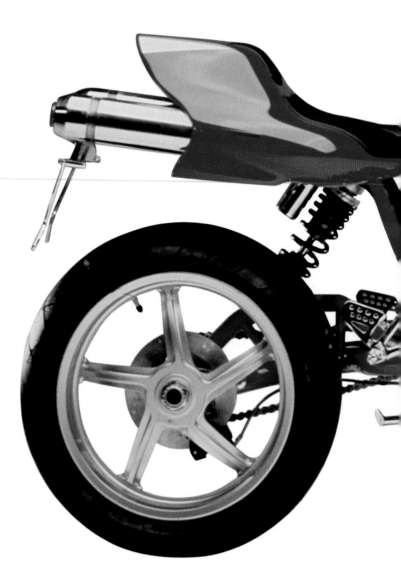

Carl Hedstrom. The last joined with George M. Hendee to create the Indian marque in 1901. Later, Indian and Harley-Davidson (1903) would dominate the vast American market.

In addition to transport, motorcycles were used for sport; apart from racing, there was growing interest in long-distance reliability trials. In 1903, the Auto Cycle Club, (now the Auto Cycle Union) was formed by several leading British motorcyclists and motorists. That same year, the club organized a ten-day, 1,609km (1,000-mile) reliability trial. This was the forerunner of the legendary International Six Days Trial (ISDT).

Then, in 1907, Brooklands, the world's first purpose-built racing circuit, was opened in the south of England. The inaugural Isle of Man Tourist Trophy (TT) race was also held that year.

With the interest in motorcycle sport came demands for more power, more flexibility and more comfort. By 1905, these had become major considerations of designers. The quest for power led to the development of the V-twin, in effect two single cylinders with a common crankcase – a 500cc single design, for example, could be used to create a 1,000cc V-twin. Some opted for four cylinders, usually, but not exclusively, of inline configuration. Not surprisingly, such developments showed up weaknesses in the transmission and chassis.

left The 2000 Ducati 900e Hailwood Evolution. Styled by South African Pierre Terblanche, it was the first motorcycle to be sold exclusively via the internet.

One notable early machine was the Autofauteuil. This featured a *fauteuil* (armchair) for sitting instead of a saddle. It had small wheels, an enclosed 427cc engine below the seat and an open frame with foot-boards. In short, it was the forerunner of the scooter, which did not become popular until almost half a century later.

In Britain, a new layout appeared for engines with two cylinders – the horizontally opposed, or flat-, twin configuration, which was employed by Douglas and, later, Germany's BMW.

By the outbreak of World War I, in 1914, the motorcycle had progressed in leaps and bounds from the experimental vehicle of only a few years before to an increasingly dependable, relatively inex-pensive and speedy means of transport. Moreover, it was gaining an ever-growing army of supporters in competitive events.

After the conflict, there was a tremendous increase in the sale and use of the motorcycle during the 1920s and 1930s, particularly in Europe and North America. Following World War II, in the late 1940s, motorcycle use became widespread around the world. As the decades have passed since, the motorcycle has gone from cheap and cheerful personal transport to a high-tech, exciting symbol of a way of life, winning back old enthusiasts and gaining new followers alike.

1 THE PION

EERS

THE PIONEERS

The earliest motorcycles were steam driven, but Gottlieb Daimler and his associate, Wilhelm Maybach, produced the first two-wheeler to be powered by an internal-combustion engine. The latter was the work of another pair of Germans, Nicolaus Otto and Eugen Langen.

The fuel Daimler used was known as benzine, but today it is called petrol, one of several distillates of petroleum, or "rock oil" (*petra*, rock; and *oleum*, oil). It is believed that the existence of petroleum and its by-products has been known for over 4,000 years, but only within the past 120 years has it served as one of mankind's most valuable commodities, providing us with unparalleled personal mobility.

One of the world's biggest priorities today is to find a suitable replacement for petrol, because it is a finite resource, but that's another story.

Daimler's creation was largely experimental, acting only as an intermediate stage in his attempt to produce what he called the "four-wheeled, passenger-carrying horseless carriage". It was left to others to further the development of the motorcycle as a form of transport in its own right.

Many pioneering designs were little more than a conventional pedal cycle with an engine attached to the frame. Although a substantial number of the original ideas that caused the birth of the motorcycle came from Germany –

together with the first series-production machines (from the
Munich-based Hildebrand & Wolfmüller factory) – the next
stage of development came from the French and Belgians.

Of particular note were the Russian-born Werner brothers in France, who did much to develop the working utility
motorcycle, popularized the diamond frame, with its central
engine position, dominated many early racing activities and
constructed the first truly practical vertical twin. The
younger Werner, Michel, died at the early age of 46, and
subsequently his brother Eugene's interest waned. When he,
too, died in 1908, the entire Werner enterprise foundered.

Englishman Edward Butler pioneered the use of the two-
stroke engine, employing the pump-type arrangement
patented by Donald Clark in 1880. In this, the injection of a
fresh charge of fuel mixture coincided with the discharge of
burnt gases on the down-stroke of the piston. On the up-
stroke, the gases were compressed, then ignited by an
electric spark to begin the next down-stroke. Butler's Velo-
cycle (three-wheeler) was exhibited as a drawing in 1884
and was not actually built until mid-1888. Thus, it was
preceded by the Daimler Einspur (one track).

GOTTLIEB DAIMLER

Today, the German engineer Gottlieb Daimler (1834–1900) is widely recognized as the "father" of the internal-combustion engine. During the latter half of the 19th century, he collaborated with Wilhelm Maybach to produce the first powered two-wheeler equipped with an internal-combustion engine.

A CRUDE DEVICE

Constructed in 1885, Daimler's machine was a crude device. It had been built as a mobile test rig for the engineer's 264 cc, single-cylinder, four-stroke powerplant. Powered by the new miracle fuel benzine (petrol), this engine had two flywheels, one on each side of the crankshaft, the whole assembly being enclosed within a cast-aluminium crankcase. "Mushroom" valves were used, the inlet being automatically operated by suction of the piston, sitting directly above the cam-operated exhaust valve in a layout known as IOE (inlet over exhaust).

The vehicle was equipped with wooden wheels and a wooden frame, an auxiliary wheel being fitted on each side. It did employ some features, however, that appeared on later motorcycle designs. The rear wheel was belt-driven, for example, while the fan-cooled engine was mounted in the frame on rubber blocks. The engine was started with a crank handle.

With a weight of 90 kg (198 lb), the Daimler machine produced 0.5 bhp at 750 rpm, while the two gear ratios provided speeds of 5 and 11 km/h (3.5 and 7 mph). Other notable aspects of the design included heated tube ignition, an evaporating carburettor and almost conventional handlebars.

THE FIRST TEST RUN

Daimler took out a patent on his design on 29 August 1885, following successful (but very short) test runs in the garden of his house and the streets of Canstatt, South West Germany; the longest journey was some 3 km (1.86 miles). He soon realized, however, that the prototype engine was not really powerful enough and his Petroleum Reitwagen, as he called it, was not easy to ride. It was not only difficult to balance, but also a real "bone-shaker", not helped by the truly dire state of the roads at the time. Consequently, Daimler concentrated his efforts on the development of what would be termed the "horseless carriage" (forerunner of the modern car), having installed one of his later engines in a coach in 1886.

Daimler's original 1885 machine can still be seen in the Daimler-Benz museum in Stuttgart-Untertufürkheim, Germany.

It should be explained that Gottlieb Daimler was a true visionary, who dreamed of his engines serving the whole of mankind. In fact, he lived to see them at work on land, water and air. Although he died three years before the Wright brothers' first powered flight at Kitty Hawk, in the USA, he did witness his engines power a balloon flight and an early motorboat, as well as a wheeled vehicle.

facing page Gottlieb Daimler. In 1885, with associate Wilhelm Maybach, he produced the first two-wheeler with an internal-combustion engine.

above By today's standards, the 1885 Daimler was very crude, having a wooden frame and steel-rimmed wooden wheels.

right The Daimler was powered by a 264 cc, single-cylinder, four-stroke engine.

HILDEBRAND & WOLFMÜLLER

The German marque Hildebrand & Wolfmüller has the distinction of being the builder of the world's first series-production motorcycle. Four men played key roles in the new venture: the brothers Heinrich and Wilhelm Hildebrand teamed up with two fellow Bavarians, Alois Wolfmüller and Hans Geisenhof, the latter having previously worked with Karl Benz. The company was formed in Munich, Bavaria, during 1894, to manufacture a new design, employing a horizontal engine mounted in a twin-tube, open duplex frame.

The Hildebrand & Wolfmüller design was unique in other ways, too. It was the first to be given the title *motorrad* (German for "motorcycle"), and it had the largest engine displacement of a series-production bike for the first 90 years of the industry. With bore and stroke dimensions of 90.0 x 117.0mm for each of its two cylinders, it displaced a whopping 1,489cc.

STEAM ENGINE INFLUENCE

The engine's pistons were linked by long connecting rods in similar fashion to those of a steam locomotive. There was no conventional flywheel; instead, the rear wheel was a solid disc and served the same purpose. Ignition was provided by a platinum hot tube, a device pioneered by Daimler, and fuel was supplied by a surface carburettor. Automatic inlet valves were fitted, while a pair of exhaust valves were operated by long rods and a cam attached to the rear wheel.

Despite its huge capacity, the engine produced only 2.5bhp, providing a top speed of almost 48km/h (30mph). This figure might seem insignificant at the beginning of the 21st century, but in 1894 it was sensational, ensuring a surge of orders to the value of over two million marks in only

left The German company of Hildebrand & Wolfmüller began manufacturing the world's first series-production motorcycle in 1894. It was powered by a massive, 1489cc, twin-cylinder engine and could achieve 48km/h (30mph). Uniquely, the role of flywheel was handled by the solid-disc rear wheel.

right The machine was not easy to start or ride, however, leading to many disgruntled customers.

a few weeks. This resulted in the company beginning work on a massive new plant in the Colosseum Strasse, Munich, able to house 1,200 employees.

In addition, a Paris-based company secured a licence to construct the Hildebrand & Wolfmüller design in France, where it was marketed under the Petrolette brand name.

PROBLEMATIC HOT-TUBE IGNITION

All was not well, however, and problems began to occur, particularly with the hot-tube ignition, which was unreliable and difficult to operate. Although the company's publicity material stated that the ignition ensured "prompt and regular explosions", the truth was somewhat different: the tubes needed to be preheated with a blowtorch before the bike could be started. Another major gripe was the poor flywheel effect

of the rear wheel, making for erratic progress.

Both German and French manufacturers also became aware quite quickly that the retail price was actually less than the manufacturing costs. The situation became even worse when the first customers began to receive their bikes. They soon began complaining bitterly, not only about poor starting, but also the difficulty of riding their machines.

The trouble came to a head when a dissatisfied customer successfully sued the French company; there followed a rush of others demanding their money back. Similar problems were experienced by the parent company in Munich.

In early 1897, both French and German companies collapsed. While the French concern returned to making pedal cycles, the German firm went into liquidation, becoming the motorcycle industry's first financial failure.

EARLY FRENCH DESIGNS

France was not only the first country to actively support motor-cycle racing, but it was also the first to have a real industry, with competition between rival manufacturers.

Well before the end of the 19th century, several French-based industrialists set up companies to market the fledgling Otto-cycle engine concept and thus created the beginnings of a viable motorcycle industry.

THE EARLY PIONEERS

Among these early pioneers were Count Albert de Dion, his mechanic Georges Bouton, Felix Théodore Millet, Ernst Michaux and Maurice Fournier. There were also the Russian born brothers Eugene and Michel Werner.

It was de Dion and the talented engineer Bouton, however, who made the biggest contribution to this new trade. Founded in Paris in 1882, de Dion-Bouton had been formed by the two men after the count had admired the workmanship of a model steam engine that Bouton had helped build.

The partnership was to give birth to the de Dion-Bouton petrol engine. Their initial power unit featured an aluminium crankcase that enclosed a crankshaft and an automatic, suction-operated inlet valve over the cam-operated exhaust valve. Unlike the German Daimler engine of the same period, which ran at 750 rpm, the French design turned over twice as rapidly, at 1,500 rpm.

The first de Dion-Bouton single was equipped with a tiny air-cooled cylinder that had bore and stroke dimensions of 50.0 x 70.0 mm, giving a capacity of 138 cc. Both cylinder head and barrel had cooling fins, while the ignition was by battery and coil, a more reliable system than the platinum hot-tube arrangement used by other manufacturers.

THE RADICAL MILLET

The most radical of all French pioneer motorcycles was the machine constructed by Felix Théodore Millet in 1892. This featured a five-cylinder radial engine fitted to the front wheel. After testing, Millet began manufacturing an improved version in 1895, which later that year took part in the world's first long-distance race. This later machine had the engine (and drive) transferred to the rear wheel.

left Peugeot was one of the pioneers of motorcycle – and automobile – design. This is a 1902 inlet-over-exhaust single with belt final drive.

above Terrot built its first motor-cycle in 1902, employing a Swiss Zedel engine. This model is a side-valve single from the late 1920s.

above By 1901, Clement was producing a machine with a 142 cc, four-stroke engine. With overhead valves, it was mounted to the frame's front downtube.

below Originally penned by Ernst Henry, the 500 cc Peugeot was an advanced design. Works rider Paul Pean is shown with the updated bike in the early 1920s.

Peugeot was another successful early French marque, but, unlike most of the others, it continued as a major concern in the two-wheel world. At the beginning of the 21st century it was still building cycles, mopeds and scooters.

In addition to constructing complete machines, Peugeot supplied other manufacturers with engines in the years leading up to World War I. For example, the winner of the first Isle of Man TT in 1907, Rem Fowler's Norton, was powered by a 726 cc Peugeot V-twin.

In 1913, Swiss engineer Ernst Henry submitted an advanced design for a brand-new racing motorcycle to Peugeot. It was powered by a 495.1 cc (62.0 x 82.0 mm) twin-cylinder engine with four valves per cylinder and gear-driven, double overhead camshafts. The engine produced 15 bhp and could propel the machine to 110 km/h (70 mph).

After the war, Peugeot commissioned the Romanian Jean Lessmann Antoinescu to update the design. This motorcycle, now with two valves and bevel-driven camshafts, gained sev-eral grand prix wins in the early 1920s. By 1923, the engine gave 27 bhp, and the bike could reach 162 km/h (100 mph).

COLLIER BROTHERS

Henry Collier and his sons, Charlie and Harry, were true pioneers of the British motorcycle industry. Like many others, they progressed to motorcycles from building pedal cycles.

The Colliers sold their machines under the Matchless name, the first prototype powered two-wheeler being constructed by Charlie and Harry in 1899, using information gleaned from a magazine article! Commercial production began in Plumstead, south-east London, during 1902, using a 2³⁄₄bhp MMC engine with automatic inlet valve (AIV).

The Colliers not only designed and built their bikes, but also rode them in competitive events with great success, winning the 1903 ACU (Auto Cycle Union) 1,000 Miles Reliability Trial.

By 1906, the range of production Matchless models stood at seven. Two were 3¹⁄₂bhp (500cc) singles; one was powered by a 427cc White & Poppe engine with T-shaped head; another had a 482cc Belgian Antoine side-valve power unit, which weighed only 63.5 kg (140 lb) fully equipped. There was a new 2¹⁄₂bhp ladies' model, powered by a 310cc AIV JAP engine, while the remaining two bikes were V-twins – one with a side-valve, 5 bhp, 745 cc Antoine unit, and the other (the top-of-the-range model) a 6 bhp, 731cc, AIV JAP.

THE FIRST TT VICTORY

On Tuesday 28 May 1907, the first ever Isle of Man TT (Tourist Trophy) race was held, having classes for singles and twins. Both Harry and Charlie Collier entered the race on 432 cc, JAP overhead-valve singles. Although Harry set the fastest lap at 67.27 km/h (41.81 mph), his younger brother won, at an average speed of 61.67 km/h (38.33 mph). This despite suffering broken front forks!

The following year, the two brothers made it a one-two in the TT – a feat still unsurpassed in the early 21st century.

Between the two world wars, Matchless built the famous 400cc Silver Hawk V-twin, followed by the even more glamorous 600cc Silver Hawk V4. In 1931, the Colliers took over the Wolverhampton-based AJS, moving production to London.

CONTINUED EXPANSION

More expansion followed in 1937 with the acquisition of the Sunbeam motorcycle marque, the group being renamed Associated Motor Cycles (AMC).

During World War II, AMC was a major supplier of military motorcycles. After the conflict, the bulk of production was exported, both Matchless and AJS bikes selling well during the early to mid-1950s, helped by participation in major racing, motocross and trials events.

James, Francis-Barnett, and, finally, Norton were also added to the group during this period. Following the death of the last Collier (Charlie) in 1954, however, a gradual decline set in, the AMC group hitting the financial "rocks" in 1966.

left Harry Collier, winner of the 1909 TT, on this 500cc Matchless single, with belt drive, magneto ignition, and spare tyres strapped to the frame just below the rider's saddle.

facing page:
top left The JAP V-twin engine that powered the 1912 Matchless Model 7.

centre 1912 Matchless 965cc V-twin with two speeds, belt final drive and acetylene lighting.

bottom left Postcard showing the company directors in 1913: H.H. Collier managing director and founder, C.R. Collier, H.A. Collier, A. Walker, and S.H. Turner.

bottom right Matchless Silver Hawk V4, built between 1930 and 1935. Its 597cc, overhead-camshaft engine gave a top-gear range from 10km/h (6mph) to over 130km/h (80mph).

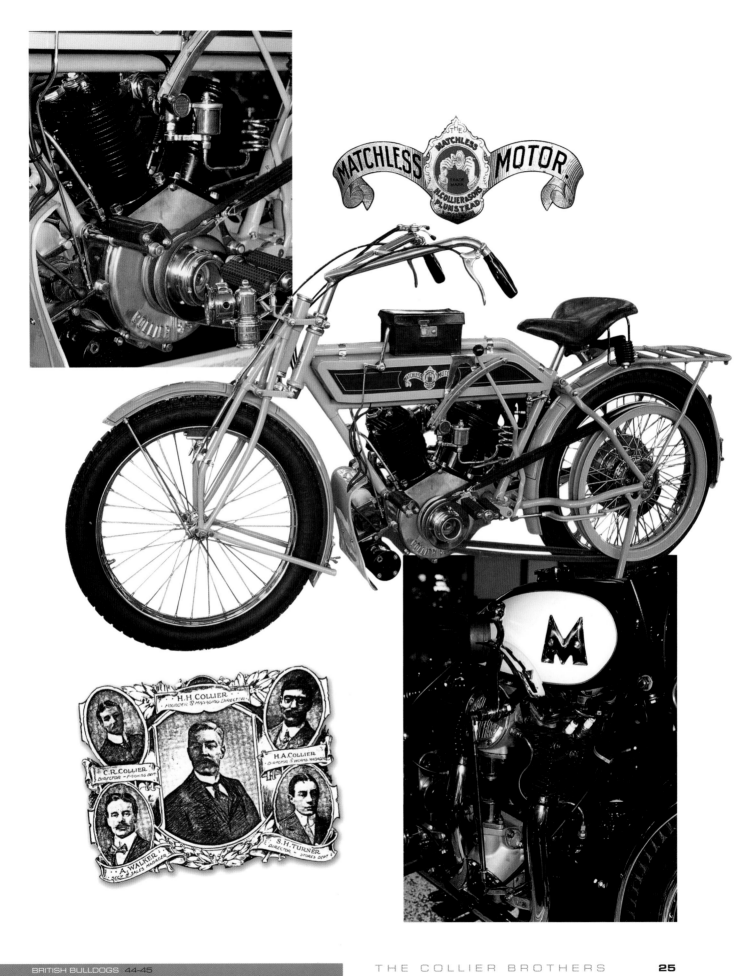

BELGIAN EFFORTS

Since World War II, Belgium has been a centre for motorcycle sporting events, including motocross and road racing – and in Spa-Francorchamps, it can boast one of the very fastest of all grand prix racing venues. The circuit is set in the wooded Ardennes area in the east of the country, only a few miles from the German border.

From the dawn of motorcycling until the outbreak of war in September 1939, however, Belgium could justly be regarded as a major motorcycle manufacturing nation, with marques of international repute, including FN, Saroléa, and, in the pioneering days, Minerva.

MINERVA

Originally founded by Sylvain de Jong to manufacture pedal cycles, Minerva co-operated later with the French company de Dion-Bouton. In 1900, de Jong purchased a Zürcher & Lüthi engine and subsequently acquired a licence to produce this 211cc, automatic-inlet-valve single. By the following year, Minerva had its own motorcycle, based on a conventional

bicycle with the engine hung from the front downtube and belt drive to the rear wheel. The company was best known, however, as a supplier of single-cylinder and V-twin engines to many pioneer firms throughout Europe. When production ceased in 1909, around 25,000 engines had been built.

FN

La Fabrique Nationale d'Armes de Guerre (FN) was set up in Herstal in 1889 to rival Saroléa as an armaments manufacturer. Like the latter, it moved into making pedal cycles before turning its attention to motorized bicycles in 1901.

The standard FN bicycle was equipped with a slender fuel tank beneath its crossbar and a 133cc engine featuring belt final drive and a "bacon-slicer" outside flywheel. Over the next three years, FN continued to develop the original design, but in 1904, the firm's chief designer, Paul Kelecom, constructed an all-new, 363cc, four-cylinder model, which was air-cooled and featured the luxury of shaft final drive. To publicize its new bike, FN organized a grand tour, taking in many of Europe's major

facing page A 1914 Belgian FN 500 with four-cylinder inline engine. Designed by Paul Kelecom, the first version, displacing 363 cc, had appeared a decade before. All featured shaft drive.

left Minerva not only produced complete motorcycles, but was also a major supplier of engine assemblies to other pioneer firms throughout Europe.

below FN was another famous Belgian manufacturer. It built its first motorcycle in 1901. This brochure dates from the early 1960s, not long before production ceased in 1965.

cities, including London and Paris. The four-cylinder engine was enlarged over the next few seasons, and by 1908 it had gained a clutch, gearbox, internal enclosed brakes, and improved spring forks.

FN continued to build and sell motorcycles until 1965.

SAROLEA

Saroléa was founded by Mainson Sarolen in Liége during 1850. Originally an armaments manufacturer, the company switched to bicycles, then entered the world of motorcycles in 1898.

Like FN and Minerva, Saroléa was heavily publicized abroad. After World War I, it entered racing; at the time, competition results influenced the sales of standard road-going models. From then until 1926, the company won countless major races throughout continental Europe, including the Belgian Grand Prix, the prestigious Liége–Nice–Liége endurance event, and the Tour of Italy.

Saroléa ceased motorcycle production in 1963.

FOUR CYLINDER

In the pioneer and vintage era, the usual way of providing customers with multi-cylinder engines was by using either a V-twin or a four-cylinder arrangement. The world's first four-cylinder motorcycle was built in 1897 by an Englishman, Major Henry Capel Lofft Holden.

THE HOLDEN DESIGN

Variable gears were well over a decade away, so Holden relied on sheer power to overcome the inherent inflexibility of direct drive. Like the German company Hildebrand & Wolfmüller, he employed long, exposed connecting rods and cranks linked directly to the rear wheel spindle. The four cylinders were placed in horizontal pairs and, with a displacement of 1,047cc, the engine operated at a mere 420rpm, producing 3bhp and giving a maximum speed of 39km/h (24mph). Production began in 1899.

The prototypes were air cooled, but production models had water cooling, which added to the already considerable weight. Consequently, it fared badly against small cars, which offered superior comfort, reliability, and price; very few were sold.

Other early four-cylinder motorcycles were produced by FN in Belgium (1904), Dürkopp in Germany (1905), Wilkinson in Britain (1908) and Laurin & Klement in Czechoslovakia (1905). Unlike the Holden, all employed inline engines.

THE AMERICAN HENDERSON

Next came the American Henderson, in 1912. Built in Detroit by Tom and William Henderson, the first of these fours dis-placed 965cc, and featured overhead inlet and side exhaust valves. Separate cylinders were mounted on a horizontally split crankcase, and the timing gears were at the front.

Proof of the robustness of the Henderson occurred when one circled the globe in 1913 – the first time this feat had been achieved on two wheels.

In 1919, William Henderson formed a new marque, Ace, to build an improved version of his four-cylinder concept. Crucially, the machine was much lighter, even though the basic design retained the inline engine in unit with a three-speed gearbox, which drove the final-drive chain. By then, the engine size had increased to 1,220cc. As was common in the USA at the time, there was only one brake, on the rear wheel.

INDIAN

Eventually, Ace was purchased by rival Indian, which continued to build the model until it launched its own four in 1928. Some 12,000 Indian fours were made, production ceasing in 1942.

In Europe, during the 1920s and 1930s, there were many four-cylinder road and racing models, from marques such as Ariel, Moto Guzzi, and Zündapp. Only two followed the inline layout, however: the Danish Nimbus (1919) and French Motobecane (1930). The former was produced by Fisker & Nielsen of Copenhagen. It was built from 1919 until 1958 and boasted several innovative features, including a massive tubular spine frame and swinging-arm rear suspension – the latter 35 years before this became an industry standard.

facing page A 1933 American Indian 1,270cc Model 433 with inline four-cylinder engine.

left The Henderson was another American four. Founded in 1912 in Detroit by Tom and William Henderson, the company was bought by Ignaz Schwinn, who added the marque to his Excelsior line.

below left The French Moto-becane concern built this 750cc, four-cylinder prototype in 1930, but it was destined never to enter production.

bottom left Famous for its razor blades, the Wilkinson company also built motorcycles in the pioneer age. This is a 1912 848cc TMC four with liquid cooling and shaft final drive.

EDOARDO BIANCHI

Born on 17 July 1865, Edoardo Bianchi can be considered the true pioneer of the Italian motorcycle industry at the beginning of the 20th century. Although he had grown up in a Milanese orphanage, this did not prevent him from becoming one of Italy's leading industrialists, thanks to a remarkable aptitude for engineering and design.

In 1885, at the age of 20, Bianchi opened a small machine shop to manufacture bicycles. Three years later, he moved to larger premises, where he produced the first Italian vehicle (a bicycle) with pneumatic tyres. Subsequently, his business expanded at a frantic pace during the remainder of the 19th century as cycling gained immense popularity in Italy, as it did throughout Europe.

In 1897, Bianchi carried out tests with an imported French de Dion-Bouton single-cylinder engine mounted in a tricycle. Although the prototype caught fire, Bianchi was justly proud in having been the first Italian to propel a road vehicle without having to resort to muscle power.

THE FIRST PROTOTYPE MOTORCYCLE

In 1901, the first prototype Bianchi motorcycle appeared, going on sale the following year. The production version was the first Bianchi vehicle to be built using components manufactured solely by the Milanese works. These included a 2 bhp engine assembly built under licence from de Dion-Bouton. A vast new factory was opened in 1902, and by 1905 Bianchi was producing not only cycles and motorcycles, but also cars.

Also in 1905, the company introduced a new design of Truffault fork. In 1910, a brand-new 500cc design made Bianchi the envy of every other motorcycle marque in Italy.

During World War I, Bianchi concentrated on aero engines, but also supplied a 647cc V-twin engine and the C75 military motorcycle. These were made in large numbers; by the end of hostilities, the V-twin had grown to 741 cc.

It was not until the 1920s that Bianchi began to enter motorcycle sport – road racing and endurance events such as the International Six Days Trial.

Bianchi

THE LEGENDARY NUVOLARI

One of Bianchi's riders was the legendary Tazio Nuvolari, who made his debut for the team in the 1925 Italian Grand Prix. The motorcycle he rode was the famous Freccia Celeste (Blue Arrow). Designed by Mario Baldi, its engine displaced 348 cc (74.0 x 81.0 mm bore and stroke) and featured double overhead camshafts, which were driven by a bevel shaft and spur gears. Unusual for the time was the enclosure of the valve gear in an oil-tight compartment. Developing 24 bhp (eventually rising to 30 bhp), the Freccia Celeste dominated the 350 cc class in Italy until the end of 1930.

The other famous Bianchi design of the inter-war years was also the work of Baldi. It was an exciting supercharged, 500 cc GP bike, which appeared at the end of 1939. With an exact capacity of 492.69 cc (52.0 x 58.0 mm bore and stroke), the four differed from other Italian designs of the period (Gilera and Benelli) in that its cylinders were vertical instead of inclined, and it relied on air for cooling instead of liquid.

ALFRED ANGUS SCOTT

Alfred Angus Scott was a visionary designer and a pioneer of the two-stroke motorcycle. His first machine, a 450cc twin, was built in 1908, the concept acting as the basis of some nine decades of two-stroke engine development. This culminated in the near domination of the small-capacity classes by this type of engine, until it was virtually killed off by legislation.

A NATURAL ENGINEERING GENIUS

Alfred Scott cultivated an engine technology that many others would pursue. His basic ideas on the two-stroke cycle endured, albeit modified as technology improved, notably from the early 1960s, when the application of acoustic theory improved gas flow, and materials were developed to withstand the increases in power that this new science provided.

Scott was a natural engineer, who originally designed his creation as a gas engine for boats, his early experiments being carried out on the River Clyde during 1900, and later in cycle frames. The Scott practice of employing a central flywheel, overhung cranks, deflector pistons, and detachable transfer ports started with these early experiments; one motor was constructed with a grooved flywheel that drove the front wheel of a conventional pedal cycle.

The specification of the 450 twin was way ahead of anything else available at the time. In the days of belt drive and bump (push) starting, the Scott excelled in having a chain drive and kickstarter, and was further enhanced by a two-speed gearbox with a central neutral position. Offering a superior performance to the rival single-cylinder four-strokes of the era, the Scott was looked on favourably by the sporting fraternity, much to the distain of other British manufacturers – who

facing page Vintage racer Chris Williams was almost unbeatable for many years with this liquid-cooled Scott. He is shown here winning at Cadwell Park in the summer of 1969.

above right Designed by Bill Cull, Scott's three-cylinder model of the 1930s had the potential to be a superbike of its era. The liquid-cooled engine featured a 120-degree crankshaft, car-type clutch and four-speed gearbox. In all, however, only eight examples were built.

right Far more typical of Alfred Angus Scott's view of two-stroke design is this Flying Squirrel, circa 1930. It has inclined, parallel twin cylinders and a radiator for the liquid cooling system above the engine.

complained bitterly to the sport's controlling body, the Auto Cycle Union. Their main objection was that, whereas their engines fired on every other revolution, the Scott fired on every revolution, which was deemed to be an unfair advantage.

AIR AND WATER COOLING

The original Scott prototype motorcycle and the early production models (1909) featured air-cooled cylinders with water-cooled heads, but soon liquid cooling was utilized for both components.

This unusual cooling arrangement was not the only pioneering feature of Scott motorcycles. In addition to the two-speed gearbox already mentioned, both the rear wheel and primary drives were by chain – at a time when belts dominated. Moreover, the fully triangulated frame was fitted

with a new design of girder front fork with a central spring, which was the forerunner of the later telescopic fork.

The roadster and racing variants of the Scott were distinguished by the colour of their cylinders – red and green respectively.

In the Isle of Man Tourist Trophy (TT) races, the Scott works racers, fitted with rotary inlet and transfer valves, broke the lap record. Wins were recorded in the 1912 and 1913 TTs, while a further lap record was gained in 1914.

The avant-garde nature of Scott's two-strokes did not totally disappear after his death in 1923. Instead, W. (Bill) Cull continued technical development, the results of which included 498- and 596cc models named the Squirrel. The larger of the two machines was capable of 137km/h (85mph) in series-production road trim.

BELT, SHAFT, OR CHAIN?

From almost the very beginning of the motorcycle, designers have had the choice of belt, chain, or (on more expensive models) shaft as a method of final drive.

THE BELT

At the very start of motorcycle manufacture, belts were the favoured option. As power outputs increased, however, so the popularity of belts waned. Many years later, though, toward the end of the 20th century, belts made something of a comeback, being employed on Harley-Davidsons, Buells, and a few Kawasaki middleweights. These later belts not only benefited from modern technology in their construction, but also had "teeth" that mated to a "sprocket", just like a chain drive. The advantage of the belt is that it does not require lubrication.

THE SHAFT

The shaft final-drive system for motorcycles was patented in 1897 by the German Alois Wolfmüller, although it was the invention of one of his employees, Ludwig Rüb. In fact, the design was based on that used on earlier Belgian FN shaft-drive bicycles. Indeed, at the beginning of the 20th century, FN became the first manufacturer successfully to specify shaft drive on its motorcycles, but it was the German BMW marque that really made shaft final drive its own.

The BMW concept, designed by Max Friz and introduced on the R32 flat-twin, was of the Cardon type (after the Italian scholar Geronimo Cardano). It featured a straight drive train, with the crankshaft, transmission shaft, and propeller shaft aligned in a row, facing the rear wheel drive.

right The majority of early motorcycles featured belt drive, direct from the engine's crankshaft to the rear wheel, as shown on this 1912 Matchless.

below right Although the German Alois Wolfmüller patented the first shaft final drive system for motorcycles, it was BMW that really established shaft drive as its own.

below The BMW R50 flat-twin employed a universal joint, connecting shaft, and bevel gears.

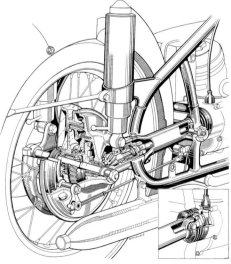

Besides high production costs, a major drawback of the shaft system was torque reaction, but in recent years this has been almost eradicated thanks to innovative design work.

THE CHAIN

The humble chain has remained the most common means of transmitting power from the gearbox to the rear wheel since the early 1920s. At first, the requisite technology came from the pedal cycle, but it was not long before stronger chains were needed to cope with ever-rising engine power outputs. Sometimes, in the very early days, block chain was used, but this only suited low-power applications, as the blocks rubbed the sprocket teeth. A version with rollers also existed, offering improved performance. While similar in appearance to the roller-bush type, it had very short outer links. These early forms were soon brushed aside, however, by the roller-type chain, which comprises an assembly of fine-tolerance components manufactured to a high standard.

PRIMARY DRIVE

Although the rear chain is usually, but not exclusively, simplex (single-row), duplex (double-row) and even triplex (triple-row) chains have been employed for the primary drive (between the engine sprocket and the clutch sprocket).

Some modern engines (from the late 1970s) have used an inverted-tooth chain for all, or part, of their primary drive. These are more commonly known as Morse or Hy-Vo chains, and are particularly long-lived.

left The 1922 250cc Radco featured all-chain drive, but the chains were exposed, leading to premature wear.

below left All-chain drive could also be fully enclosed, as on this 1915 AJS. This protected both chains and rider.

2 BIKES FO

R WAR

above The German BMW company pioneered the use of hydraulically damped telescopic front forks, as fitted to this 1935 R12 750.

left The military began using motorcycles in World War I (1914–18). These machines are Royal Enfield V-twins.

facing page An Indian Chief 1200 cc, side-valve V-twin on patrol near the front, circa 1942. This machine was typical of the large-capacity American motorcycles used by Allied forces during World War II.

BIKES FOR WAR

In the military, the motorcycle replaced the horse as an efficient means of carrying despatches, while in sidecar guise, it proved a functional and highly mobile light assault vehicle and transport.

It was during World War I (1914–18) that the motorcycle first displayed its suitability as a war machine, the British forces with their Douglas 350s, Triumph 550s, Rudge Multis, P & Ms, and Royal Enfields leading the way. Following America's entry into the conflict, a host of Excelsiors, Harley-Davidsons, and Indians were shipped across the Atlantic. Other early military motorcycles came from FN (Belgium), NSU (Germany), and Puch (Austro-Hungary).

When Germany invaded Poland on 1 September 1939,

at the start of World War II, many countries in Europe had been gearing up for war for several months, following the Munich Crisis of the previous year. The British Norton company, for example, had foregone its usual challenge for road racing honours during the 1939 season, concentrating instead on military contracts, thanks to the persistence of its managing director, Gilbert Smith.

The machine Norton was building was the 490 cc, side-valve, single-cylinder Model 16H. Basically, this was a 1937 civilian model, the military conversion consisting of little more than a crankcase shield, a pillion seat or rear carrier rack, a pair of canvas pannier bags, provision for masked lighting, and an overall coat of khaki paint. Norton's factory

in Birmingham also built the very similar Big Four with a larger 634 cc engine. Around 100,000 16Hs were built for wartime service, demonstrating how Smith's foresight had paid commercial dividends.

Like the Nortons, many other military motorcycles were little more than revamped civilian models. In modern warfare, however, as in other areas, the development of the motorcycle was unrelenting, and soon more specialized designs began to appear.

Each country's armed forces found they had particular requirements: the British concentrated on simple singles; the Germans favoured complex, horizontally opposed twins; the Americans opted for heavyweight V-twins. The Italians

produced some interesting designs, many of them built by companies that were more familiar with grand prix racing than military needs.

In many ways, the motorcycle was the unsung hero of World War II. It carried out a truly amazing variety of tasks, but was used most widely by the famous DRs (despatch riders). Although radio communications had replaced the miles of telephone wire that had been laid near the front lines during World War I, the motorcycle played a vital role in carrying messages and documents.

Other important duties included marshalling road convoys, scouting, and, when coupled to a sidecar, acting as a machine-gun carrier.

AMERICA FLIES THE FLAG

Although America did not officially enter World War II until the Japanese attack at Pearl Harbour in December 1941, within days of Britain and France declaring war on Germany, in September 1939, both governments had placed substantial orders with the American motorcycle industry.

During the conflict, two companies supplied the bulk of the American motorcycle output for European armies, and later for the American and Canadian forces. These were Harley-Davidson and Indian.

To European eyes, the American machines were massive (and heavy!). Most were V-twins with foot-operated clutches and hand-change gearboxes.

HARLEY-DAVIDSON WLA

Essentially, the Harley-Davidson WLA was the civilian 745cc (69.8x96.8mm bore and stroke), side-valve V-twin with a number of modifications to make it suitable for military service. These included a heavy-duty luggage carrier, a rifle scabbard, "blackout" lights at front and rear, and a shield to protect the underside of the engine's crankcase. The military specification also included leather pannier bags, front and rear crash bars, a handlebar-mounted screen, and, of course, an overall coat of drab paint. Other features were a comprehensive air filtration system, footboards, and, quite often, ammunition boxes, which were mounted on the front forks.

The engine was of the traditional Harley-Davidson narrow–angle V-type. If anything, this was over-engineered, the iron cylinders and light-alloy heads being provided with more than adequate finning to ensure sufficient cooling, even during long periods of low-speed running in bottom gear, when acting as convoy escort, for example.

The valves were on the offside (right) of the power unit, the two exhaust header pipes being joined low down to run rearward into a single silencer. Other details included a single Linkert carburettor, coil ignition, a three-speed gearbox, and dry-sump lubrication.

The WLC (Canadian) version featured a hand-operated clutch instead of a foot clutch and hand gearchange. Around 20,000 WLCs were produced, compared with 68,000 WLAs. In Harley "speak", the WL was known as the Flathead 45.

INDIAN 841

Although there were several conventional Indian V-twins (very much in the mould of the Harley-Davidson WLA), an interesting alternative was the Model 841, which had been conceived specifically as a military motorcycle. This was a 745cc (73.0 x 88.9mm bore and stroke), side-valve, transverse V-twin. Compared with other American designs of the period, it was an advanced machine. As well as state-of-the-art suspension (telescopic front fork/plunger rear), the 841 featured a four-speed, foot-change gearbox instead of the hand-change type favoured by Harley-Davidson. It also had shaft final drive. Although transverse V-twins were nothing new, like the later Moto Guzzi V7 series, the Indian offered excellent cooling, a low centre of gravity, and good performance, with a top speed of 113km/h (70mph).

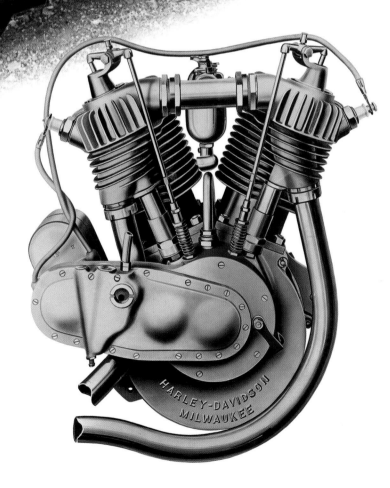

facing page An American military training college during 1917, with instructor, pupils, and Harley-Davidson V-twin engine.

above During World War II, Harley-Davidson built almost 90,000 WLA (American) and WLC (Canadian) 750cc V-twins for military use.

right Harley-Davidson, inlet-over-exhaust V-twin engine, circa 1917.

AMERICA FLIES THE FLAG

THE FATHERLAND

Long before World War II, the Nazi Party was hard at work promoting Germany in motorsport – Auto Union and Mercedes-Benz on four wheels; DKW, NSU, and BMW on two. With typical Teutonic thoroughness, each of the bike builders would concentrate on a certain class in grand prix events: DKW, 250 cc; NSU, 350 cc; and BMW, 500 cc. In practice, however, the supercharged, double-overhead-camshaft NSU twin proved a disappointment, so DKW ended up also competing in the 350 cc category.

At the same time, many German "civilian" motorsport organizations had required their members to train on military equipment for several years before war was finally declared.

A RATIONALIZED INDUSTRY

The German motorcycle industry was rationalized in 1938 by Colonel von Schell, who had been granted complete control in a manner that was only possible under a totalitarian regime. The number of powered two-wheelers was slashed from 150 to 30. This left small two-strokes from Ardie, DKW, TWN, and

Zündapp, while Victoria produced a four-stroke. NSU made the 250 OS and 601 OSL models – both with overhead-valve, single-cylinder engines and separate four-speed gearboxes – plus the extraordinary Kettenkrad. This last machine was a strange combination of tracked vehicle and motorcycle, being powered by a liquid-cooled, 1,478 cc, overhead-valve, Opel four-cylinder engine.

It was the massive BMW and Zündapp horizontal twins, however, that became the definitive German military motorcycles of the era.

BMW R75

The R75 (like its Zündapp counterpart, the KS750) was a superbly built machine, but it was very expensive to manufacture (twice as costly as a VW jeep!). From the start, it was designed as a sidecar outfit, having a power take-off (with lockable differential) to drive the sidecar wheel.

Another interesting technical feature was the gearbox, which offered four-speeds and reverse, and was connected to

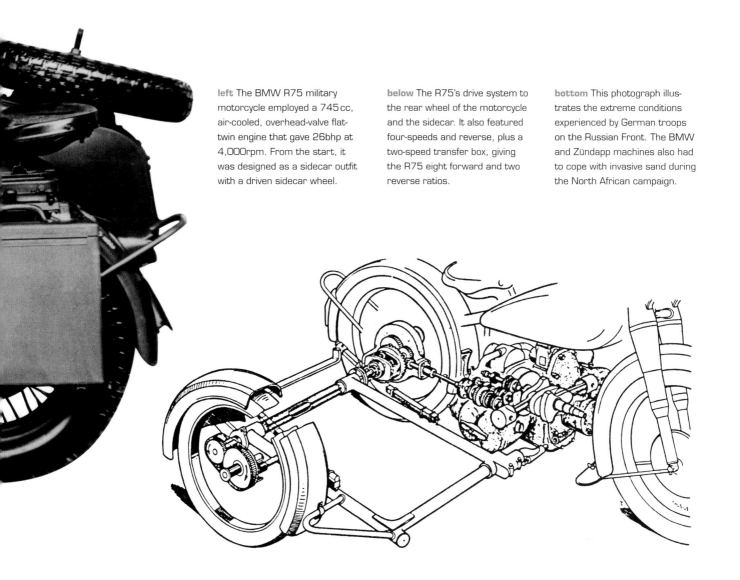

left The BMW R75 military motorcycle employed a 745cc, air-cooled, overhead-valve flat-twin engine that gave 26bhp at 4,000rpm. From the start, it was designed as a sidecar outfit with a driven sidecar wheel.

below The R75's drive system to the rear wheel of the motorcycle and the sidecar. It also featured four-speeds and reverse, plus a two-speed transfer box, giving the R75 eight forward and two reverse ratios.

bottom This photograph illustrates the extreme conditions experienced by German troops on the Russian Front. The BMW and Zündapp machines also had to cope with invasive sand during the North African campaign.

a two-speed transfer box, giving the R75 a total of eight forward and two reverse gears.

At 420kg (924lb), the R75 was a heavy machine, and even with hydraulic operation, the brakes struggled under adverse conditions. The 745cc (78.0 x 78.0mm bore and stroke), air-cooled, overhead-valve flat-twin engine delivered 26bhp at 4,000rpm, and with a compression ratio of 5.6:1, it was extraordinarily tractable. In all, 16,500 were built.

ZÜNDAPP KS750W

The flat-twin Zündapp, with its unusual chain-and-sprocket gearbox, had been around since the early 1930s, the military KS750W appearing in North Africa in autumn 1940. Originally, it had been intended as a towing vehicle for a light gun for airborne troops, but its front wheel would lift clear of the ground when used for this purpose! So it was adapted as a multi-terrain machine. Displacing 751cc, the K750 had a pressed-steel frame, telescopic front fork, sidecar wheel drive, and large-section 400mm (16in) tyres.

BRITISH BULLDOGS

The Allies were victorious in World War II, but which nation's bikes were the most successful in wartime conditions? If quantity is the yardstick, then Great Britain has to be awarded the accolade – British manufacturers produced more military motorcycles during the war than those of any other nation, the most popular types being BSA (126,000, mainly side-valve M20s), Norton (100,000, side-valve 16Hs), AMC (80,000, mostly Matchless G3/Ls), Ariel (47,600, overhead-valve W-NGs), and Royal Enfield (29,000, two-stroke WLs). If the machines supplied by Triumph, Velocette, Cotton, Douglas, James, and Excelsior are included, the total output reaches 420,000 – a staggering figure.

MATCHLESS G3/L

Of all the British four-stroke models, the best all-round machine was the Matchless G3/L. Like the German DKW

RT125, it was a classic design and, in civilian guise, it remained in production until the mid-1960s.

Unlike the German lightweight, however, the Matchless was based on an existing model, the G3. Both had a capacity of 348cc (69.0x93.0mm bore and stroke) and a four-speed, Burman-made, foot-change gearbox.

What really set the G3/L apart from its earlier civilian counterpart was its weight – the military model had been pruned significantly and was 25kg (56lb) lighter.

Most significant of all – at least, as far as British bikes were concerned – was the incorporation of the patented Teledraulic front fork, a design that relied heavily on pre-war BMW technology. This was not only technically superior to the girder types found on other British machinery of the period, but it also improved the bike's aesthetic appearance, giving the newcomer a very modern look.

The air-cooled, overhead-valve single, with a vertical cylinder, alloy head, coil valve springs, gear-driven cams, and roller-bearing big-end, produced 16bhp and could reach a top speed of 114 km/h (71 mph).

The machine's only real drawback – when compared with the more bulky (and much heavier) Norton and BSA side-valve singles – was that, because of its sheer compactness, gaining access to some components was difficult and time consuming. Around 60,000 G3/Ls were built for military service.

AIRBORNE BIKES

British airborne forces employed lightweight motorcycles that could be dropped by parachute where needed and quickly assembled in the field. Although others were built in Italy, Germany, and America (all in small numbers), the British had

by far the greatest success with the concept. Specialized machines were built by James, Royal Enfield and Excelsior, all being two-strokes.

Royal Enfield's effort (known as the Flying Flea) was a 126cc (54.0x55.0mm bore and stroke), DKW-based, ultra-lightweight model designed by Ted Pardoe. The James was powered by a 122cc (50.0x62.0mm bore and stroke), Villiers 9D engine with three-speed, hand-change gearbox, and was known as the ML (Military Lightweight).

The most unusual of the trio, however, was Excelsior's folding design, known as the Welbike. This had been designed by Lieutenant-Colonel J.R.V. Dolphin, and was equipped with tiny wire wheels and a 98cc, single-speed, push-start Villiers engine. The machine could be stowed in a standard cylindrical parachute container. After the war, it was fitted with an Excelsior engine and sold as the Corgi.

DKW RT125

Although Alfred Scott was the original pioneer of the two-stroke engine, German companies – first DKW and later MZ – were largely responsible for its evolution.

JÖRGE SKAFTE RASMUSSEN

The founding father of DKW was Jörge Skafte Rasmussen. Born in Denmark in July 1898, he moved with his family to Dusseldorf, Germany, in 1904, then in 1907 to Zschopau, south of Chemnitz in Saxony. Later, he held a number of engineering posts and, after World War I, formed J.S. Rasmussen, the forerunner of DKW.

In 1916, inspired by acute wartime fuel shortages, Rasmussen had begun to experiment with a steam car, the Dampf Kraft Wagen. Although the machine was not put into production, it provided the now-famous DKW initials.

Three years later, in 1919, Rasmussen produced an 18 cc, toy two-stroke engine, together with a 122cc auxiliary engine, both designed by Hugo Ruppe.

WORLD'S LARGEST MOTORCYCLE MANUFACTURER

Rasmussen progressed to building motorized bicycles, "armchair" motorcycles (forerunners of the modern scooter), and, finally, conventional motorcycles. By 1926, DKW had manufactured 100,000 engines (all of them two-strokes), while the motorcycle range included bikes with displacements of 200, 250, 350, and 500cc.

By the time DKW became part of Auto Union, in 1932, it could justifiably claim to be the largest motorcycle manufacturer in the world.

RACING IMPROVES THE BREED

DKW had begun racing two-stroke motorcycles in 1925, the machines having intercooling and the Bichrome system of supercharging. In 1931, chief designer Hermann Weber, assisted by August Prussing, created a split-single design, and from then on until the outbreak of war in 1939, DKW was a dominant force in not only grand prix racing, but also record-breaking in the lightweight (250cc) class.

THE RT125

The company had the largest research and development department in the industry, and all of the technical expertise and experimentation had a knock-on effect on production models. The prime example of this was DKW's most famous and influential bike, the RT125. Designed by Weber, this tiny motorcycle was powered by an all-new, 122cc (52.0 x 58.0 mm bore and stroke), unit-construction engine with a light-alloy cylinder head and cast-iron barrel.

Despite the rationalization of the German motorcycle industry, carried out in 1938, the RT125 was such an excellent design that it was ordered into production as soon as its test programme had been completed. Its light weight and durability led to its wide use by the German armed forces throughout World War II, particularly for scouting and communications work.

After the end of the war, the RT125 went on to become the most copied motorcycle in history. The British BSA Bantam, American Harley-Davidson two-stroke and Soviet Moska all stemmed from Weber's design.

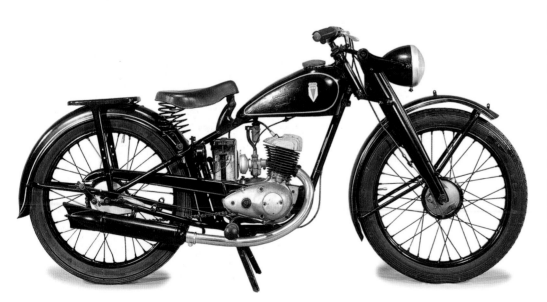

left Powered by an all-new, 122 cc, piston-port-induction, two-stroke engine with integral three-speed gearbox, the DKW RT125 was the work of the brilliant designer Hermann Weber.

facing page:
top In October 1937, DKW took part in a successful series of record-breaking attempts. Both solo and "sidecar" (shown) versions were used.

bottom Ewald Kluge and Siegfried Wünsche rebuilding a DKW 250cc racing engine. The German government was very keen to support motorcycle sport in the years leading up to World War II.

DKW RT125 SPECIFICATIONS

Engine Air-cooled, piston-port, two-stroke, single-cylinder

Displacement 122 cc

Bore and stroke 52.0 x 58.0 mm

Ignition Flywheel magneto

Gearbox Three-speed, foot-change

Final drive Chain

Frame All-steel, welded

Dry weight 68 kg (150 lb)

Maximum power 4.8 bhp at 5,000 rpm

Top speed 76 km/h (47 mph)

ITALIAN MACHINES

Military bikes from Italy carried marque names that were better known on the race circuit than the battlefield: Benelli, Bianchi, Gilera, and Moto Guzzi. The pick of the designs were the Gilera Marte, with a 500cc, side-valve engine and shaft final drive, and the Moto Guzzi Alce, with its typical Guzzi horizontal, single-cylinder unit. The Italians also took their *motocarri* (motorcycle truck) to war, notably the Gilera Gigante VT in 500 and 600cc guises, and Guzzi's Trialce design.

GILERA MARTE

The Gilera Marte was a purpose-built military motorcycle. Its engine was based on the existing 498cc (84.0 x 90.0mm bore and stroke), side-valve LTE single, but with an alloy head, higher compression and revised timing to produce 14bhp at 4,800rpm. Much of the balance of the machine was new, however, and technically interesting.

There was a significant change in the transmission, where gears replaced a chain for the primary drive; the final drive was through a pair of bevel gears to a shaft that transmitted power to the rear wheel. A bevel box drove the rear wheel from the crown wheel and included a spur gear set, which drove a cross-shaft ahead of the rear axle. A second set of spur gears at the sidecar wheel (most Martes were built as sidecar machines) transferred the drive back again so the two rear wheels were in line. The sidecar gears also included a dog clutch, which was controlled by a hand lever on the motorcycle, to engage the drive as required.

While the frame remained much as the civilian LTE, the offside (right) rear fork was totally new and took the form of a bell-crank. This was made from steel pressings welded together. The sidecar wheel was suspended by a trailing arm, the "chair" being mounted on the right, so the drive from the rear wheel matched, also being on the right.

This layout meant that Gilera was the only manufacturer during World War II to offer a sidecar outfit where both the wheels of the bike and sidecar had their own suspension systems – as well as having drive to both wheels. Even BMW and Zündapp failed to match the Italian design in this area – the German motorcycles had rigidly mounted rear wheels. Thus, the Marte was unique among its military counterparts during the conflict.

MOTO GUZZI ALCE

Of all the Italian motorcycles that went to war, the Guzzi Alce and Trialce were by far the most widely used. Compared with the Gilera Marte, however, both were less technically advanced. In fact, the solo Alce was obsolete when it entered service in 1939, with its ancient inlet-over-exhaust valve operation for the 498cc (88.0 x 82.0mm bore and stroke), horizontal, single-cylinder engine.

The Trialce, fitted with the same power unit, was a much more interesting creation, having twin driven rear wheels. These machines were often attached to a sidecar for a host of duties, ranging from machine-gun platform to bread carrier.

Both the Alce and Trialce had been developed from the earlier, but very similar, GT17 (1932–39).

facing page A line of Moto Guzzi 500cc singles with traditional horizontal cylinders outside Addis Ababa, April 1941. The Guzzi was built in larger numbers than any other Italian motorcycle during World War II.

above The Gilera Marte was powered by a 498cc, side-valve, single-cylinder engine. It was unique in that all three wheels had their own suspension systems. The sidecar wheel was also driven.

left Moto Guzzi's definitive military motorcycle of the 1930s, the GT17, was pressed into service at home and abroad. Some were equipped with a Breda heavy machine-gun, as shown.

FRONT FORKS

Over the years, there have been many different designs of motorcycle front fork, the most popular types being girder, trailing-link, leading-link, Earles, and telescopic. The last, employed by BMW and Matchless during World War II, became standard equipment on just about every machine manufactured since the war; only scooters, mopeds, and limited-production "specials" have used other types.

THE TELESCOPIC CONCEPT

The concept of the telescopic fork, where the wheel axis moves in a straight line parallel to the steering head, is not new. In fact, as early as 1910, the British Scott was using it. Other manufacturers followed suit, but it was not until 1935 that the modern-style telescopic fork with hydraulic damping appeared on two BMW models. Today, telescopic forks come in all sizes, both conventional and the latest upside-down (inverted) type.

Some types have rebound-only or two-way damping, or no damping, the last reserved almost exclusively for small, commuter-type machines. There is never a total lack of damping, however, as the friction generated between the moving parts of the fork provides some, although this is kept to the bare minimum, as it affects the fork's operation.

Designers of modern motorcycles go to considerable lengths to reduce the "stiction" (natural resistance to movement) found in the forks, as invariably this is greater than the friction once the motorcycle is moving. By employing the latest materials and coatings, these forces can be minimized.

DESIGN FEATURES

A telescopic fork comprises a pair of legs (or sliders), to which the wheel is fixed, and a pair of tubes (known as stanchions), which are clamped in a pair of yokes. In most cases, the springs are housed in the tubes with the damping mechanism. The tubes are normally hard-chromed for longer life and appearance, and each leg will have one or more oil seals. The legs, into which the tubes are fitted, are manufactured from either aluminium or steel. On really exotic machines (mainly for racing), magnesium has also been used for fork legs.

As machines have become more powerful, it has become necessary not only to increase the size and strength of the various fork components, but also to reinforce the legs with either mudguard supports or separate braces.

The acceleration possible with the latest powerful bikes applies substantial forces to the fork, which must be capable of withstanding them. Deceleration places even greater loads on the front suspension, however, particularly when braking hard from high speeds. During the 1980s, anti-dive systems were seen as a way of combating these forces, but in the end proved of little use. Similarly, replacements for the telescopic fork, such as hub steering, have been offered, but none has found favour.

3 THE 1920

0S & 1930S

above left Designed by Val Page, the 647 cc Triumph 6/1 vertical twin preceded Edward Turner's famous Speed Twin by several years. It was introduced in July 1933.

left Folke Mannerstedt returned home to Sweden to join Husqvarna from the Belgian FN company in 1928. One of his first tasks was to build this new 500 cc racer powered by a British JAP engine.

above The British Excelsior company won the 1933 Lightweight (250cc) TT with its innovative Mechanical Marvel. The bike featured four radial valves, two carburettors, and a sophisticated dry-sump lubrication system.

facing page A typical British motorcycle of the era, a 1930 twin-port Ariel Model G with 499 cc, overhead-valve engine, hand-change gearbox, rigid frame, girder front fork, and drum brakes.

THE 1920S & 1930S

After World War I (1914–18), the population of Europe had a much better understanding of the mechanical world. At the beginning of the conflict, cavalry had gone to war on horseback; by the end, horses had been replaced by motorized armoured tanks. In virtually every other military arm, the development of machinery had taken place at breathtaking speed.

Throughout Europe, large numbers of new companies were formed to cater for peacetime motorcycling. Some survived for only a few months, whereas others joined the existing motorcycle industry to become major players.

Although motorcycles were manufactured in Czechoslovakia, France, Belgium, Sweden, and Switzerland, the major producers resided in Germany, Italy, Britain, and the United States of America.

Many of the war's largest armaments concerns attempted to grab a slice of the action, but their efforts were often hampered by the damage their factories had received during the conflict. Probably the nearest to success was the British Sopwith company's involvement with the ABC 400 cc flat-twin. Although this was an advanced design, with such features as overhead valves and a four-speed gearbox, it suffered from insufficient development, and both ABC and Sopwith went bust during the 1920s.

Much more successful was the German aero engine giant BMW. At first, it only produced engines for other

motorcycle firms, but in 1923, the first machine to bear the famous blue and white emblem on its tank appeared at the Paris motorcycle show.

The first truly golden age of motorcycle sales was the 1920s. Then came New York's Wall Street Crash in October 1929. Although America bore the early brunt of this, it was not long before the "Great Depression", as it was called, spread across the Atlantic to Europe, millions being thrown out of work in all the industrialized countries.

The effects of the Great Depression lasted until the mid-1930s, forcing many firms out of business, while others only escaped by the skin of their teeth.

There were several important advances in motorcycle design during the 1920s and 1930s, many the work of notable engineers. Among these were Edward Turner, Valentine Page, Percy Goodman, Arthur Carroll, Walter Moore, George Patchett, James L. Norton, and Harold Willis (Britain); Mario Mazzetti, Alfonso Morini, Adalberto Garelli, Carlo Guzzi, Mario Baldi, Piero Remor, and Carlo Gianni (Italy); Max Fitz, Richard Küchen, Hermann Reeb, Hugo Ruppe, August Prussing, and Hermann Weber (Germany); and Charles B. Franklin and Bill Harley (USA).

After World War II (1939–45), several important manufacturers did not re-open their factories, notably Brough Superior, Böhmerland, Coventry Eagle, Della Ferrera, New Imperial, OK Supreme, and Rudge.

THE SWEDISH INDUSTRY

Although there have been many Swedish motorcycle marques over the years (including Nordstjernan, Nymans, and Monark), one company has dominated – Husqvarna.

The Husqvarna company was formed as early as 1699, when Erik Dahlberg, a leading industrialist and governor of the region in which the town of Huskvarna (note the "k" in place of the "q") was situated, decided to supply much-needed weapons to the Swedish Army. Thus, like BSA in Britain, Husqvarna built its original business upon the arms trade.

With its expertise in precision engineering, Husqvarna was quick to seize the opportunity when the motor age arrived, and in 1903 it began building motorcycles. Its first machines were fitted with imported engines from Moto Rêve, NSU, and FN.

A COMPLETE MOTORCYCLE

The next major step came in 1918, when Husqvarna designed and built its own complete machines for the first time. A large contract from the Swedish Army followed, and the motorcycle side of the business became an important money spinner. It

was then that the company made its first moves into motor-cycle sport.

Realizing that competition wins were the finest form of publicity, Husqvarna developed special bikes for long-distance cross-country events, then popular throughout Scandinavia. These, in turn, led to entries in the International Six Days Trial.

The young Swedish engineer Folke Mannerstedt joined the company, having returned from Belgium, where he had worked for FN. One of his first tasks was to design a new racer, which was powered by a British JAP 500 cc, overhead-valve, single-cylinder engine. After indifferent results, however, Mannerstedt, assisted by Calle Heimdahl, set out to build a new engine.

THE V-TWIN

Mannerstedt's first prototype appeared at the 1930 Swedish Grand Prix, and from 1932, it became a serious challenger. The engine was a 497 cc (65.0 x 75.0 mm bore and stroke), 50-degree V-twin with pushrod-operated valves. It produced 44 bhp at 6,800 rpm and was good for 193 km/h (120 mph).

By 1934, it had been improved still further, with larger valves and a stiffened crankcase, resulting in 46 bhp. There was also a 350cc version.

The engine was fairly conventional in that the crankcase was separated vertically, with pushrod tubes on the right side and exposed rockers. Two carburettors were squeezed between the cylinders, angled to give a down-draught effect.

TWO-STROKES

At the end of 1935, Husqvarna quit the GP scene. In the same year, the company built its first two-stroke, a commuter bike powered by a 98cc engine with two-speed gearbox. This set a precedent for postwar production.

The two-strokes grew in capacity, and by 1960 the focus of attention was on off-road competition, notably motocross. During the following decade, the marque captured ten world motocross titles. The machine that provided the basis for this success was the 22bhp, 250cc model, developed from the 175cc, three-speed road machines of the mid-1950s.

top left The most famous and demanding of all Swedish motorcycle trials was the Novemberkåsan (November Trophy). This photograph shows the start of the 1916 event.

top During the early and mid-1930s, Husqvarna successfully raced a 497cc, 50-degree V-twin designed by Folke Mannerstedt. These three examples are shown at the 1932 Swedish Grand Prix.

above This machine was one of the first Husqvarna models. It was equipped with an imported Belgian FN single-cylinder engine, being built between 1903 and 1908.

JAMES LANSDOWNE NORTON

Born in Birmingham in 1869, James Lansdowne Norton was the son of a cabinet-maker. His talents, however, lay elsewhere. Anything mechanical fascinated him, and his practical abilities were remarkable – at the age of 12, he constructed a working steam engine. On leaving school at 15, he was apprenticed as a toolmaker and soon became involved in the manufacture of cycle chains. In 1898, he established his first business, supplying components to the bicycle trade.

THE FIRST MOTORCYCLE

The first true Norton motorcycle appeared in November 1902. Called the Engerette, it was basically a conventional pedal cycle with a 143cc Clement engine "clipped" to the front downtube of the frame. Later, there was a larger-engined version, powered by a Swiss Moto Rêve V-twin.

Norton tested his products by competing in speed trials and reliability runs. In 1907, he made his big breakthrough when a Peugeot-engined Norton V-twin, ridden by Rem Fowler, won the twin-cylinder class of the inaugural Isle of Man TT.

As the first decade of the 20th century came to an end, "Pa" Norton decided to race in the TT himself – a surprising decision, as he was then over 40 years old and in ill health. Unfortunately, his debut ended in retirement, but he did break new ground by becoming the first rider to pilot a Norton single around the TT course.

As a result of competition successes, Norton coined the advertising slogan, "The Unapproachable Norton". With the help of such men as racer, record breaker, and tuner Daniel O'Donovan, he built on that success, Norton machines also doing well at the Brooklands track, which had opened in 1907.

THE MODEL 18

By the beginning of the 1920s, Norton's range of side-valve machines had begun to stagnate, but in 1921 the first overhead-valve engine appeared. This retained the marque's traditional long-stroke, 79.0 x 100.0 mm dimensions and displaced 490 cc. It was used to power the Model 18, which was introduced in 1922 and built for many years. A Model 18, ridden by Alec Bennett, gave Norton its second TT victory, in 1924.

His health having deteriorated, James Norton passed away on 21 April 1925. However, he had laid the foundations for the greatness that was to follow – during 1924, Nortons had come first in no fewer than 120 sporting events.

The marque went on to become one of the most successful in the world, a fitting tribute to its remarkable founder, James Lansdowne Norton.

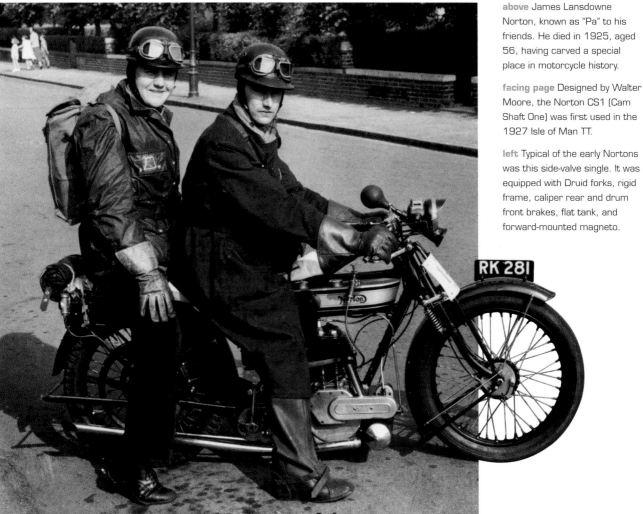

above James Lansdowne Norton, known as "Pa" to his friends. He died in 1925, aged 56, having carved a special place in motorcycle history.

facing page Designed by Walter Moore, the Norton CS1 (Cam Shaft One) was first used in the 1927 Isle of Man TT.

left Typical of the early Nortons was this side-valve single. It was equipped with Druid forks, rigid frame, caliper rear and drum front brakes, flat tank, and forward-mounted magneto.

CARLO GUZZI

Carlo Guzzi (1889–1964) was not only founder of Moto Guzzi, but also one of the truly great designers of all time. For many years, Moto Guzzi stood for highly innovative, trend-setting engineering. This was epitomized by Guzzi's first motorcycle, which appeared in 1919.

The machine had been conceived during World War I by Carlo Guzzi, a mechanic, and two pilots, Giovanni Ravelli and Giorgio Parodi. The trio were ardent motorcycle enthusiasts who agreed to pool their talents and resources after the war to create a company to manufacture Guzzi's brainchild. Sadly, Ravelli was killed in a flying accident during the last days of the conflict, but Guzzi and Parodi survived to fulfil their dream and later used a flying emblem logo to remember their fallen comrade. The bike carried the legend GP (Guzzi-Parodi) on its tank.

UNORTHODOX FEATURES

In Guzzi's prototype, unorthodox features abounded, many having been taken directly from aeronautical engineering practice. A horizontal cylinder gave improved cooling and greater stability through its lower centre of gravity. The engine's bore and stroke dimensions were over-square at 88.0 x 82.0mm, providing a lower piston speed and greater volumetric efficiency. Four valves contributed to superior breathing and were operated by a single overhead camshaft, driven by a shaft and bevel gears. An external flywheel not only

offered better balance, but also allowed the use of a much smaller crankcase. Maximum power was 17bhp, providing a top speed of around 129 km/h (80mph) – outstanding figures for a single-cylinder 500 at the time.

PRODUCTION BEGINS

By the time production began in 1921 (under the Moto Guzzi name), some features had been dropped to save costs (the engine no longer had an overhead camshaft, and there were only two valves). Carlo Guzzi had demonstrated, however, that he was a creative and forward-thinking engineer, something he would do time and again over the following years.

Moto Guzzi went racing thanks to publicist Giorgio Parodi, who correctly perceived that, as a newcomer, the quickest way for the marque to establish itself in the industry was to be seen in competition.

BREADTH OF DESIGN

Moto Guzzi racing machines displayed an unequalled breadth of design: horizontal singles, parallel twins, V-twins (inline and transverse), three- and four-cylinder models (horizontal and inline), and, the most amazing of all and still unmatched, the fabulous V8 of the 1950s.

Although Carlo Guzzi died in 1964, the marque he helped found is still going strong today.

MOTO GUZZI

facing page The classic big single Guzzi, the Falcone. This employed two overhead valves and a horizontal cylinder, with bore and stroke dimensions of 88.0x82.0mm.

left Carlo Guzzi (standing) with his 492cc, across-the-frame, four-cylinder grand prix racer at Monza in June 1931.

below Other features of the Falcone single included magneto ignition, a dynamo and a foot-change gearbox with typical Italian heel-and-toe operation.

BMW R32

BMW's first stab at a motorcycle engine was not a success. The work of Kurt Hanfland, it was a 148cc, two-stroke single, marketed under the name Flink in 1920. The Flink was followed, in 1921, by a far more substantial engine, which effectively was the true beginning of the BMW two-wheel story. Designed by Martin Stolle and designated M2B15, it was a 493cc, flat-twin side-valve with bore and stroke dimensions of 58.0x68.0mm. It was supplied to the likes of Bison, SBD, SMW, and Victoria.

In 1922, the M2B15 was used by BMW itself to power the Helios motorcycle, the engine being mounted fore-and-aft (like the British Douglas) and driving the rear wheel by chain. The Helios chassis was not a BMW design, however, even though production took place at the firm's Munich works. Unfortunately, the machine was not a very good motorcycle, a fact that led BMW's chief engineer, Max Friz (who had been responsible for the company's excellent aero engines), to put aside his personal dislike of two-wheelers and set out to create a new machine, one that BMW could be proud of.

ENTER THE R32

When the results of Friz's efforts were unveiled at the Paris Salon in October 1923, the bike caused a sensation. Its engine was mounted transversely as a single unit with a three-speed gearbox and shaft final drive, while the frame was a full twin-triangle design. The front fork was sprung by a quarter-elliptic leaf spring. It was the first BMW to carry the now-famous blue and white circular emblem (the quartering of this logo portrayed a whirling aircraft propeller and resembled Bavaria's chequered flag).

Although the R32 was not as powerful as some contemporary machines, it was superior in several important aspects. It offered a truly modern motorcycle concept at a time when unreliable engines, flimsy frames, and temperamental transmissions were commonplace.

Key features of the revolutionary design were its sprung front wheel (an industry first), its flat-twin engine, installed transversely to provide equal cooling for both cylinders, and its shaft drive to the rear wheel. Wisely, Friz decided not to

far left The first BMW motorcycle to sport the famous badge, the R32 flat-twin, designed by Max Friz, made a sensational debut at the Paris Salon in October 1923.

left Production of the new R32 at BMW's Munich works, around 1924. Significantly, the work was carried out on benches, not a production line.

below R32 rider Fritz Bieber on his way to winning the 1924 German national 500cc road racing championship.

employ bevel gears to turn the drive from the crankshaft through 90 degrees, as would have been necessary with a conventional chain-drive arrangement. Instead, he opted for the simpler and mechanically more reliable shaft, which took the drive directly to the rear wheel, where a pinion meshed with a ring gear in the wheel hub.

Extremely effective, the R32 inspired subsequent generations of BMW motorcycle design. Even in the early 21st century, the same basic flat-twin engine configuration and shaft drive are employed by the German company – testament to the enduring nature of Friz's 1923 machine.

BMW R32 SPECIFICATIONS

Engine Air-cooled, side-valve, transverse, flat-twin

Displacement 494cc

Bore and stroke 68.0x68.0mm

Ignition Magneto

Gearbox Three-speed, hand-change

Final drive Shaft

Frame All-steel, welded

Dry weight 120kg (264 lb)

Maximum power 8.5bhp at 3,200rpm

Top speed 90km/h (56mph)

ARIEL SQUARE FOUR

By 1930, four-cylinder motorcycles were no great innovation, examples already having been produced in Britain, the USA, Germany, Denmark, and Belgium. Even so, the appearance of the Ariel Square Four at London's Olympia show that year made a huge impact because of its compact size, neat appearance, low weight and complete freedom from engine vibration.

A SQUARE FORMAT

Previously, there had been inline fours and V4s, but the square layout of the Ariel's cylinders was without precedent. The machine had been conceived by Edward Turner, the original prototype having an engine size of 498cc. Besides its unique layout, the engine featured a chain-driven, single overhead camshaft, and overhung crankshaft flywheels for all four cylinders (subsequently changed to overhung cranks on only three cylinders). There were two cranks, each having a large helical gear cut on the centre flywheel. This expensive method was changed to spur gears from the 200th engine.

HORIZONTALLY SPLIT CRANKCASES

Unusually for a British design, the Square Four's crankcases were split horizontally, the engine's main bearing being clamped in the top section. The separate cylinder head and one-piece cylinder block were cast in iron. A single carburettor was mounted between the two forward-facing exhaust header pipes, while an induction passage ran to all four cylinders.

The compactness of the engine and four-speed, hand-change, Burman gearbox is indicated by the fact that they were fitted to the existing Ariel 500cc Sloper single chassis.

Series production of the 500 began in 1931, a larger 601cc model being added in 1932. This was achieved simply by increasing the bore size of the cylinders. From 1933, the 500 was available only to special order.

THE 4G

In 1937, a new, larger model, the 4G, was introduced. This featured a 997cc, overhead-valve engine with vertically split crankcases, dry-sump lubrication and plain-bearing big-ends. In place of the overhung cranks, there were two forged crankshafts, each having a central flange that carried a separate flywheel. The carburettor was rear-mounted, and there was a foot-change gearbox.

Production of the 4G was halted during World War II, but the machine reappeared in mid-1945, with girder forks and plunger rear suspension designed by Frank Anstey (earlier models had had rigid frames). Telescopic front forks were introduced in mid-1946.

For 1949, the engine received an aluminium top-end, which, together with other weight-saving measures, led to a weight decrease of 14.9kg (33lb). Magneto ignition was replaced by a coil-and-battery arrangement with a car-type, four-cylinder distributor.

The four-exhaust-pipe 4G Mark II was introduced in 1953, having separate exhaust manifolds and revised styling. The final update occurred for the 1956 season, the machine receiving a cowled headlamp, a full-width alloy front brake, increased oil tank capacity and a duplex, endless timing chain. Production came to an end in 1959.

left A 1939 Ariel catalogue, showing the 4G 1000, with 997cc, overhead-valve engine and four-speed, foot-change gearbox, which had been introduced in 1937.

facing page:
top A 1953 4G Mark II four-pipe Square Four, with all-alloy engine, separate exhaust header pipes, plunger frame, and telescopic front forks.

bottom Early versions of the Square Four, such as this 1932 601cc model, employed an overhead camshaft, iron head and barrel, rear-mounted magneto, twin exhaust pipes, and hand-operated, three-speed gearbox.

1937 ARIEL 4G 1000 SPECIFICATIONS

Engine Air-cooled, overhead-valve, square-four

Displacement 997cc

Bore and stroke 65.0 x 75.0mm

Ignition Magneto

Gearbox Four-speed, foot-change

Final drive Chain

Frame All-steel, single front downtube

Dry weight 191kg (420lb)

Maximum power 36bhp at 6,000rpm

Top speed 153km/h (95mph)

THE USA: BIKE V. CAR

In the early years of the 20th century, the USA had a large and diverse motorcycle industry; machines from both Harley-Davidson and Indian were exported in considerable numbers to Europe. Moreover, both marques did well in racing on both sides of the Atlantic. In fact, by 1915, Indian was able to claim to be the largest manufacturer in the world, having sold a record 31,950 motorcycles that year. It is interesting to note that Indian was more advanced than Harley-Davidson in the technical stakes.

In the years leading up to World War I, the competition between Indian and Harley-Davidson produced rapid technical improvements, and not just in engines and frames. There were significant advances in transmission systems, with gearboxes and clutches being introduced.

The war helped Harley-Davidson more than Indian, however. The former also gained considerable publicity from its famous Wrecking Crew racing team, which, from 1914 until 1921, achieved a dominant position in the American domestic racing scene.

COMPETITION FROM FOUR WHEELS

After World War I, the American motorcycle industry found itself fighting another battle – against the increasing popularity of the car, particularly in the shape of Henry Ford's Model T. The main attraction of the Model T was its bargain price, a consequence of the innovative production-line techniques used by Ford in his Detroit plant. By 1918, the Model T had already been on sale for a decade, but its real impact was felt between

that year and 1927, after which it was discontinued in favour of more modern types.

Although there was a vast increase in motorcycle sales in Europe (nearly 100 new marques appeared in Britain alone) during this period, the huge upsurge of car production in the USA, headed by the Model T, slashed the previous large number of American bike producers. Only the fittest and largest survived, marques such as Harley-Davidson, Indian, Excelsior, and Henderson.

1,200 CUBIC CENTIMETRES

From a total production of 28,189 motorcycles in 1920, Harley-Davidson's output slumped to just over 10,000 the following year. But the company didn't take this lying down and responded by launching a new, 1,200 cc, 45-degree V-twin.

As for Indian, its legendary designer Charles B. Franklin penned the 600cc Scout, which appeared in late 1919, followed by the 1,000cc Chief in 1922. A year later, Indian introduced the 1,200cc Big Chief.

All of these designs were side-valve V-twins, however, and although they were basically sound machines, they often cost as much as a car and didn't change much from year to year.

Indian was purchased by E. Paul du Pont in 1930 (after suffering financial problems), while both Excelsior and Henderson folded in 1931. Quite simply, the American motorcycle industry was being left behind by its European cousin. Although both Harley-Davidson and Indian went on to survive until the post-World War II period, Indian closed in 1953.

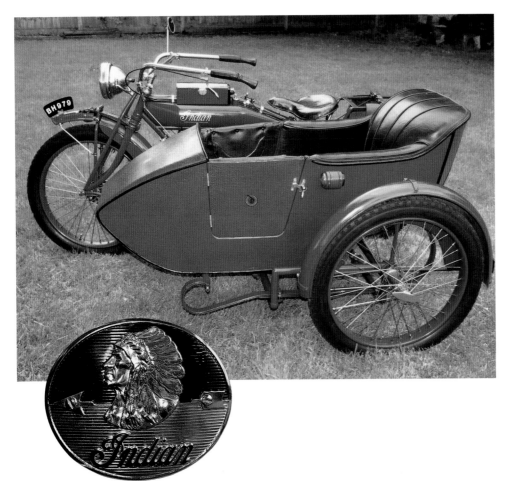

facing page During the early years of the 20th century, the USA had a substantial and diverse motorcycle industry. Machines from the likes of Harley-Davidson (a 1914 C10 single is seen here) and Indian were exported to Europe in considerable quantities.

above left The motorcycle was widely used for police duties during the 1920s. This photograph dates from 1924 and shows how officers in Los Angeles were able to arrest offenders on the spot.

above Harley-Davidson was also very active in motorcycle sport before and immediately after World War I. Alberto Trivellato is seen here with the 1000cc V-twin he used to win the Merluzzi Hill Climb, circa 1919.

left An early 1920s Indian V-twin and sidecar.

BROUGH SUPERIOR

Often referred to as "the Rolls Royce of motorcycles", the Brough Superior owed its existence to George Brough's father, William. The latter's foresight and enthusiasm had led to the formation of the famous Nottingham-based company.

The first all-Brough-engineered motorcycle appeared in 1902 and, thanks to its outstanding performance and handling, several examples achieved success in reliability trials. It was in this arena that George first came to prominence, riding many of the bikes to victory.

WHICH TWIN?

Despite the success of Brough machines, there was considerable disagreement between father and son in respect of the engine configuration. In 1913, George had been seen aboard a V-twin, which thereafter he favoured, whereas William Brough preferred the flat-twin (like Douglas and later BMW). In the early days, William's opinion ruled, resulting in flat-twin Broughs, but this brought George into open conflict with his father over future developments. Consequently, he set up in business on his own account, using the name Brough Superior. Quite what his father thought of this situation is best left unrecorded!

George Brough's idea was to use the very best components and technology available at the time, the results being some of the most respected motorcycles ever built. His creations were not only fast and reliable, but also good-looking and of high quality, thanks to such features as the famous Brough nickel-plated tank.

far left The Brough Superior (an SS100 is seen here) was not only a very fast roadster, but also a successful racing and record-breaking machine between the world wars.

left During the late 1930s, Brough built a small number of four-cylinder Golden Dreams.

below Creator George Brough with the very first of the SS100 Pendine models.

SS100

In 1924, George Brough produced his most well-known model, the legendary SS100. Each of these machines was sold with a certificate to confirm that it had reached 160 km/h (100mph) on the test track.

Basically, the SS100 was a refinement of the existing twin-camshaft, double-valve-spring SS80. It was introduced with a 988cc (85.5x86.0mm bore and stroke), overhead-valve, JAP V-twin engine and three-speed, hand-change gearbox.

The SS100 rapidly captured the public's imagination, and development continued. For 1926, the Alpine GS version was introduced, with a 995cc (80.0x99.0mm bore and stroke) JAP engine featuring triple valve springs and more power. There was also the Pendine, guaranteed to have been tested

at no less than 176km/h (110mph) and named after the speed record venue on the south coast of Wales.

From 1928, rear suspension was available; but traditionalists preferred the rigid frame, as it was simpler and lighter, providing better control of the rear wheel.

A new SS100 appeared at the end of 1932. This had a JAP V-twin tuned to produce a claimed 74bhp at 6,200rpm and 200km/h (125mph). The specification included twin carburettors, twin magnetos and a selection of compression ratios, from 6:1 to 12:1 (the latter for racing only).

By the mid-1930s, a positive-stop, foot-change gearbox replaced the hand-change type; a year later, a four-speed Norton gearbox arrived. Finally, from 1936, the 50-degree, 990cc, Matchless V-twin replaced the JAP engine.

ADALBERTO GARELLI

Mention Garelli to most enthusiasts and the word "moped" would probably spring to their minds. The marque's founder, however, Adalberto Garelli, was one of the pioneers of the Italian motorcycle industry.

Garelli was born in Turin in 1886, and at the age of 22 he gained a degree in engineering, subsequently dedicating his work to the study and perfection of the first two-stroke engine at Fiat. The automobile giant did not share his enthusiasm for two-strokes, however, and he left the company in 1911.

THE SPLIT-SINGLE CONCEPT

In 1912, Garelli came up with the concept of a two-stroke split-single. It consisted of two cylinders cast in a single block, with a common combustion chamber, the two pistons working in parallel, both connected by a long gudgeon pin to a single connecting rod. Each piston had a capacity of 174.6cc, making a total of 349cc. The bore and stroke dimensions were 50.0x89.0mm.

To prove his concept, in 1914 Garelli set out to climb

Moncenisio Pass in the Alps. In succeeding, he became the first to achieve this feat on a motorcycle.

That year, Garelli also joined Bianchi as head of its motorcycle division. His next stop was Stucchi, another important motorcycle factory of the period.

During this time, he won a competition organized by the Italian Army for a military motorcycle, using a special version of his 349cc split-single. Finally, in 1919 he was able to realize a personal dream and set up his own plant, at Sesto San Giovanni, near Milan.

A VICTORIOUS DEBUT

Almost from the outset, Garelli's design attracted attention and admiration, thanks in no small part to a victorious debut in the Milano-Napoli road race. This was a remarkable success for a manufacturer that had only just begun production.

The race took place over a 840km (522-mile) route linking two of Italy's major cities. In those days, the roads were mostly of hard earth, not asphalt, but, for all that, the distance was

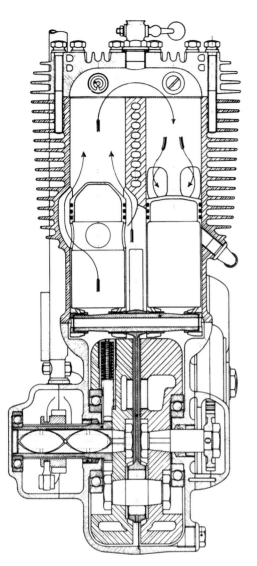

facing page Adalberto Garelli dedicated his life to the study and perfection of the two-stroke motorcycle engine.

left The Garelli split-single engine, showing the two pistons and their single, long connecting rod.

above A typical 1970s Garelli moped, the single-speed automatic Eureka.

below The front cover of *Moto Ciclismo*, dated 14 September 1922, showing Garelli rider Ernesto Gnesa after his victory in that month's Italian Grand Prix.

covered in 21 hours, 56 minutes, at an average speed of 38.27 km/h (23.79 mph); the rider was Ettore Girardi.

Many more victories followed over the ensuing few years; Garelli was also successful in several speed record attempts.

SEPARATE LUBRICATION

One of Garelli's 350cc racers was the first to be equipped with an oil tank for separate engine lubrication. Like the most sophisticated modern two-strokes, the oil was mixed automatically with the fuel, in a quantity that was proportioned accurately by the throttle setting. Another milestone in two-stroke technology was the use of an expansion-chamber exhaust, which used the shape of the exhaust to maximize power.

The final 1926 model of Garelli's racing 350 split-single developed 20 bhp at 4,500 rpm and could reach a speed of 142 km/h (88 mph). It featured a two-speed, hand-change gearbox. The oil/petrol ratio was 17 percent, and the machine weighed in (dry) at 98 kg (216 lb).

NSU: TEUTONIC MASTERCLASS

Although not the first, NSU (Neckarsulmer Strickwaren Union) was one of the real pioneering marques of the German motorcycle. It also became one of the biggest and best in that country. During the first half of the 20th century, NSU often led the world in design, innovation, and production methods, to say nothing of its sporting successes.

A WORKING PROTOTYPE

NSU was formed originally, in 1873, to manufacture and repair knitting machines. In 1900, however, it produced a prototype motorcycle equipped with a single-cylinder engine. This entered production the following year. Although sturdy, the machine was crude, its Swiss-made, Zedel clip-on engine being mounted at an incline from the frame's front downtube. Engine power was transmitted to the back wheel by a direct-drive belt, although a conventional bicycle pedal crank and chain were retained to allow the machine to be pedalled if the engine needed assistance.

Before long, NSU switched to engines of its own design, including a range of V-twins with capacities ranging from 496 to 996 cc.

THE PRODUCTION LINE

The company pioneered production-line techniques for the German industry. It also took the export market seriously, opening a sales office in London during 1905.

Almost from the beginning, NSU recognized the importance of motorcycle sport as a proving ground for its technol-ogy and as a marketing tool for its products. Not only did it go racing from 1905, but it also made many successful speed record attempts over succeeding years.

Although production was switched to munitions during World War I, the firm rapidly returned to motorcycles after the conflict. By 1922, the works was operating at full capacity, with over 3,000 on the payroll.

In 1929, the famous British designer Walter Moore left Norton and joined NSU. This caused considerable controversy, his designs for the German company often being referred to as "Norton Spares Used".

TWO-STROKES

Many of Moore's NSU designs were completely original, however. One of his most successful was the 98 cc, two-stroke Quick autocycle. Between 1936 and 1953, almost a quarter of a million examples were manufactured. Besides a full range of motorcycles, with capacities from 97 to 592 cc, NSU built the Motosulm, essentially a moped with a 63 cc, single-cylinder engine mounted over the front wheel, which it drove by chain.

The various four-stroke NSUs of the 1930s were built in side-valve, overhead-valve and overhead-camshaft forms. All were noted for their high standard of engineering and superb finish, which was often better than BMW's.

During the 1950s, NSU sold millions of machines. The company also broke more world speed records – culminating in 338 km/h (210 mph) in 1956 – and gained world champion-ships in the 125 and 250 cc categories (1953–55).

facing page The company soon began building engines of its own design, including a range of V-twins.

above Overhead-camshaft 500 and 600cc singles were designed by Walter Moore in the mid-1930s.

right A 1938 supercharged, 345cc, double-overhead-camshaft racing twin.

below During the 1950s, NSU built the overhead-camshaft, 247cc Max and Supermax touring bikes.

below right A 1910 NSU V-twin with early rear suspension.

FOOT-CHANGE GEARBOX

In the very early years of the motorcycle, drive was taken directly from the engine shaft to the driven wheel, usually by means of a belt. As a result, there was no way of isolating the wheel from the shocks and irregularities of the torque delivered by the engine. At first, manufacturers looked at improving the flexibility and smoothness of the engine, hence the comparatively large number of multi-cylinder engines introduced so early in the history of the motorcycle. Twin- and four-cylinder examples began to appear almost as soon as the motorcycle was invented.

PEDAL ASSISTED

Nevertheless, the problems of intractability remained, and the belt did nothing to diminish them. This was also a reason why many early machines were equipped with pedals, which could be used when starting off, stopping, climbing hills and the like.

Then, during the first decade of the 20th century, two-speed transmissions appeared in the de Dion-Bouton, Fafnir, NSU, Royal Enfield, and Scott, while Chater-Lea introduced a three-speed gearbox as early as 1906. There was also a variable-ratio pulley system devised by Rudge-Whitworth for its famous Rudge Multi model (a similar system was used on the Zenith Gradua).

THE BIG BREAKTHROUGH

While all-chain drive (primary and final), primary gears, and shaft final drive came as by-products of the gearbox and clutch mechanisms, there still remained the relatively slow and complicated (with levers, rods, and joints) hand-operation. For a long time, control of the gearbox was by a hand lever, which moved through a quadrant to provide a progressive sequence of gear selection. The rider's hands were already pretty well occupied, however, so it was clearly desirable to transfer some of these duties to his feet. Using a pedal to change gear was

an attractive concept, but it was not compatible with the quadrant type of progressive change, which required more sensitivity than a heavily booted foot could achieve.

The problem was solved by Velocette chief designer Harold Willis in 1928, when his positive-stop mechanism was introduced on the works TT model. It soon became standard on Velocette's production KTT racer.

With the positive-stop system, the gear lever no longer moved to a separate position for each gear, but was simply prodded to engage the next highest or lowest gear. Then, it would spring back to its original position.

Soon, other manufacturers began employing positive-stop operation. Later, four-speed gearboxes replaced the two- and three-speed types. Five, and sometimes six, ratios arrived in the 1960s, having first been used in racing. In the 1970s, Honda and Moto Guzzi introduced models with automatic gear-boxes, but these proved unpopular and were soon dropped.

TRIUMPH SPEED TWIN

Edward Turner's Speed Twin was one of the truly great motorcycle designs of the 20th century. Launched at the end of July 1937, it not only propelled Triumph into the big league, but also made the traditional big singles and V-twins seem old hat.

A SIMPLE DESIGN

As with all great designs, the Speed Twin was essentially simple. Amazingly, the engine was actually lighter and narrower than the Tiger 90 single of the same era. Turner could have built a flat-twin (like BMW and Douglas) or a V-twin (like JAP and Matchless), but chose a vertical layout because it offered the prospect of the best power-to-weight ratio, the layout making it possible to design a compact assembly. Other reasons, as Turner explained, included "…even firing intervals, perfect cooling, and extreme rigidity, which is the essence of high power output; the engine will go into a smaller frame without loss of accessibility; the two cylinders do precisely the same amount of work, and the engine is amenable to all single-cylinder tuning techniques… This type of engine is superior to any other form of twin except on balance."

OVERHEAD VALVES

At first, Turner had planned an overhead-camshaft layout, but this was soon abandoned in favour of a twin-camshaft, pushrod design with overhead valves. The arrangement was pleasingly symmetrical, the gear-driven camshafts being positioned fore and aft of the cylinder barrels.

A valuable aspect of Edward Turner's design genius was his ability to utilize existing components to minimize production costs. A good example of this was the 498.76cc displacement of the new Speed Twin, which was based on identical 63.0x80.0mm bore and stroke dimensions to Triumph's Tiger 70 250 single. Thus, production could be rationalized by employing the same pistons, rings, small-ends, and circlips.

ORIGINAL FEATURES

Even so, several features of the new Triumph engine were original. One was the patented built-up crankshaft, which had no middle bearing. There were also light-alloy connecting rods. Another patented construction was the big-end. This was unusual in several respects. Each cylinder's con-rod bore directly on its crankpin, but the steel bearing cap featured a thin lining of white metal. This arrangement took advantage of the high heat conductivity of the alloy, while the white-metalled cap acted as a safety feature because, in the event of oil loss, the metal would melt, preventing seizure.

Coded 5T, the Speed Twin was built until the late 1950s in its original form, with separate gearbox. A unit-construction (combined engine and gearbox) version was introduced, as the 5TA, for 1959; it was offered until the end of 1966.

1937 TRIUMPH SPEED TWIN
SPECIFICATIONS

Engine	Air-cooled, overhead-valve, vertical twin
Displacement	499cc
Bore and stroke	63.0x80.0mm
Ignition	Magneto
Gearbox	Four-speed, foot-change
Final drive	Chain
Frame	All-steel, tubular construction
Dry weight	160kg (353lb)
Maximum power	26bhp at 6,000rpm
Top speed	145km/h (90mph)

facing page A 1949 Triumph advertisement, proclaiming the virtues of the Speed Twin and the fact that the design had a ten-year start over its rivals.

left A 1946 5T Speed Twin, with front-mounted dynamo, the first postwar Triumph.

below Edward Turner's Speed Twin was one of the truly great motorcycles of the 20th century. Launched in July 1937, it started a fashion for vertical twins that other manufacturers would follow.

4 THE POSTWAR BOOM

above left Almost 25,000 German Dürkopp Diana scooters were built between 1954 and 1960. Its single-cylinder, two-stroke engine displaced 194cc.

left Introduced in 1950, the German Horex Regina was powered by a 342cc, four-stroke single with pushrod-operated valves and twin exhaust ports.

above During the immediate postwar period, motorcycles were widely used not only for commuting, but also for touring, as these Italians prove during a visit to Britain on their 250cc Comet in 1956.

right A 1950s British Triumph 500cc Speed Twin, capturing the carefree spirit of the day.

THE POSTWAR BOOM

After the end of World War II in 1945, the world's motorcycle manufacturers enjoyed a sales boom that lasted until the end of the 1950s.

Initially, the world market was split into three distinct areas: Continental Europe, North America, and Britain. In Europe, the emphasis was on cheap transport, and there was an enlightened attitude to mopeds – no test or insurance was required for machines under 50cc with pedals. This helped establish a two-wheel culture based on mopeds, scooters, and lightweight motorcycles. In North America, the population was wealthier, and many large-capacity roadsters and dirt bikes were bought as fun

vehicles. In Britain, the economy was in such a poor state that for the rest of the 1940s, every machine produced had to go for export to help pay off the nation's debts incurred during six long years of war.

Peak motorcycle sales were achieved during the 1950s, when many new designs were introduced, notably from Germany, Italy, and Britain. Manufacturers in Spain, France, and Sweden also built relatively large numbers of machines. The American domestic motorcycle industry stagnated, however, and Harley-Davidson was the only company that managed to survive in the long term.

There were several important developments during the

1945–59 period. Besides the growth in moped and scooter sales, a number of new 500, 650, and 700cc, twin-cylinder models were introduced by British manufacturers. Technical advances included suspension improvements (the telescopic front fork and swinging-arm rear with twin hydraulically damped shock absorbers), the dualseat, 12-volt electrics, and push-button starting on some German designs.

There was also a vibrant motorcycle industry behind the Iron Curtain. East Germany produced IFA (later MZ) and Simson machines; Czechoslovakia, Jawa, and CZ; Poland, WSK, and Junak. The Soviet Union turned out copies of pre-war German DKW and BMW designs.

Although everything seemed fine in the world motor-cycle market, there were problems, and they were serious. For a start, the founding fathers of the industry – the pioneers – men such as the Collier brothers, Carlo Guzzi, and Edward Turner were growing old. In addition, steadily rising standards of living toward the end of the 1950s meant that, for the first time, many potential customers could afford a small car. Then there was the growing challenge from the Far East, where the Japanese had been working hard to build a vast industry, which would provide serious competition for the West in the 1960s.

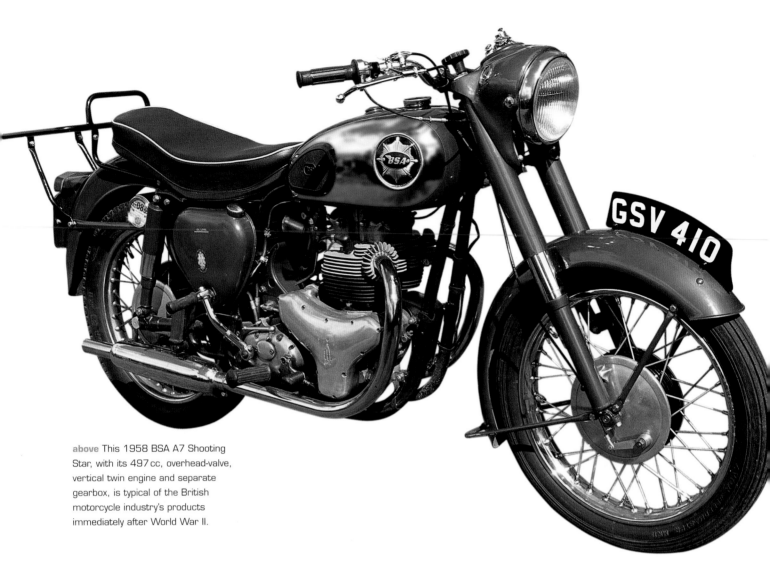

above This 1958 BSA A7 Shooting Star, with its 497 cc, overhead-valve, vertical twin engine and separate gearbox, is typical of the British motorcycle industry's products immediately after World War II.

BRITISH INDUSTRY

Although the British had been on the winning side, when World War II ended in 1945, the nation was at a low ebb financially; the years of conflict had taken a fearsome toll on the country's economy. Thus, even though there was a ready-made home market waiting to buy its products, the British motorcycle industry was forced to sell everything it could produce overseas, prompted by Sir Stafford Cripps' call to British industry in his 1945 budget speech: "Export or bust!"

RAISING TARGETS

Where motorcycles were concerned, Cripps had demanded that, by 1948, export figures be raised to a staggering 258 percent of the 1938 level. The manufacturers responded by achieving 278 percent by 1947. Some industry suppliers also took part in this export drive, including Lucas (electrical equipment), Smiths (instruments), and Amal (carburettors).

The main overseas markets were the USA, Canada, Australia, and Europe. In fact, it would be true to say that such names as BSA, Norton, and Triumph became as well-known

abroad as at home during the immediate postwar period. Triumph, in particular, carved a reputation in North America to rival the domestic Harley-Davidson and Indian brands.

THE MARQUES

At the time, Britain seemed to have an almost endless list of manufacturers. The major companies were AMC (AJS and Matchless), the BSA Group (BSA, Triumph, Sunbeam, and Ariel), Douglas, Norton, Panther, Royal Enfield, Velocette, and Vincent. There were scores of smaller marques, too, such as Cotton, DMW, DOT, Excelsior, Francis-Barnett, James, New Hudson, Scott, and Sun.

BEST-SELLERS

BSA not only controlled the most powerful grouping, but at that time was also the best-selling brand in the world. From its giant plant in Small Heath, Birmingham, came its established B31 and B33 singles, together with new twins in the shape of the 497 cc A7 and 646 cc A10 models. Later, the company

left P & M (Panther) was a Yorkshire firm that had built its first motorcycle in 1904. The company's machines are seen here at an exhibition in Copenhagen, Denmark, in 1951.

below left Matchless was based in south-east London and belonged to the AMC (Associated Motor Cycles) Group. By 1957, this also included the AJS, Norton, James, and Francis-Barnett marques.

below A police-specification Matchless G12 646cc twin on patrol with London's Metropolitan force, circa 1962.

bottom During the 1950s, the Automobile Association's sturdy, side-valve BSA M21 and box sidecar outfits were common sights on British roads.

developed high-performance versions, including the Gold Star, Shooting Star, Road Rocket, Super Rocket, and, finally, the Rocket Gold Star.

Another member of the group, Triumph, had introduced the trend-setting Speed Twin model in the late 1930s, which was joined in late 1949 by the 649cc Thunderbird. Like their BSA counterparts, there were high-performance versions of these machines, among them the Tiger 100, Tiger 110, and T120 Bonneville.

Ariel (also a major brand within the BSA Group) built both 500 and 650cc vertical twins, as did rivals Norton, AJS, Matchless, and Royal Enfield. All of these marques produced large numbers of single-cylinder machines as well, often updated prewar models.

Most British motorcycles of the 1940s and 1950s featured separate engines and gearboxes. From the beginning of the 1960s, however, some firms, notably BSA and Triumph, switched to unit construction, where the engine, gearbox, and clutch were combined in a single assembly.

TRIUMPH SPEED TWIN 76–77
CAFE RACER CULTURE 108–109

THE SCOOTER AGE

Although the first scooters appeared shortly after World War I, the golden era for sales ran from the mid-1940s to the mid-1960s, when they met a demand for inexpensive personal transport. At the beginning of the 21st century, however, a new breed of these small-wheeled machines began to appear, this time aimed at providing mobility with style.

THE ITALIANS SET THE PACE

Immediately after World War II, the Italian aircraft maker Piaggio and engineering giant Innocenti almost put their entire country back on wheels with their respective Vespa and Lambretta scooters.

Prohibited from building aircraft, in 1945 Piaggio created a small monocoque machine. Its design was so advanced that its shape and engineering principles would form the basis of all Vespa scooters sold over the next 50 years or so. Originally displacing 98cc, it evolved through 125 and 150cc versions before being given a 200cc engine. Vespa never deviated from its original concept of a monocoque body shell with the engine mounted inside and adjacent to the rear wheel.

By contrast, the Lambretta retained its tubular frame, having either shaft or fully enclosed chain final drive.

DOUGLAS AND NSU

The vast sales enjoyed by the Vespa and Lambretta soon attracted attention abroad and, in 1949, the Bristol-based Douglas company announced its intention to build a version of the former. The British company made a number of changes

to the Piaggio design, one of which was the relocation of the headlamp from its original position on the front mudguard to the front apron, to meet UK law.

In Germany, NSU purchased a licence to build the Lambretta LC in 1951. This had been the Italian marque's first model to feature enclosed bodywork, having been introduced a year earlier.

When NSU's licence expired, the company created the Prima series, the first model being launched at the Brussels motorcycle show in January 1956. Although, at first glance, the Prima D appeared similar to the Lambretta LC, in fact, it was much improved. It was a true luxury scooter, with an electric starter and a new, hydraulic, rear shock absorber.

THE GERMAN INDUSTRY RESPONDS

While the British industry never really followed Douglas's lead to mass-produce scooters, the Germans did. After NSU's Prima came a host of innovative luxury models, including the TWN Contessa, Maico Mobil, and Maicoletta, Dürkopp Diana, Heinkel Tourist, and Zündapp Bella.

Heinkel had also been an aircraft manufacturer, and in 1952, unable to continue in this field, it began building the Tourist. This had a 174cc, overhead-valve, four-stroke, fan-cooled engine. It was quiet, powerful, and comfortable. A tubular frame supported steel body panels and cast-aluminium footboards, while the engine was rubber mounted to eliminate vibration. As one commentator put it at the time, "If the Heinkel were a car, it would truly be deserving of executive status."

left Lambretta and rival Vespa were the most successful of all scooter marques. This is Lambretta's LD150, built between 1954 and 1958.

facing page:
top Like Piaggio, Heinkel put its aviation engineering expertise into scooters. The result was the high-quality Tourist.

bottom left The 1959 Iso Milano had a 150cc engine.

bottom right Zündapp, one of Germany's premier motorcycle marques, built the Bella scooter. The R203 had a 197cc, single-cylinder, two-stroke engine and could reach 93km/h (58mph).

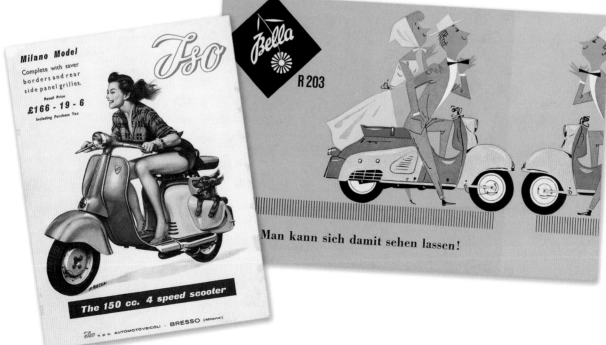

VICTORIA
Avanti

The stylish sports moped and light weight motorcycle

Les vélomoteurs sportifs de rasse et motocyclettes légères

El gallardo velomotor deportivo y la arrojada motocicleta ligera

bohringer

MOPEDS

Most early motorcycles were motorized bicycles and, in the 1920s, this simple form of machine was developed to become the autocycle, the forerunner of the moped. Unlike the true moped, which invariably had an engine of 50cc or less, the autocycle was usually around 100cc.

CLIP-ONS

Immediately after World War II, the "clip-on" appeared, being an engine that was simply clipped to a conventional pedal cycle frame. Leading manufacturers included Garelli, Power Pak, and Ducati, the last being the most interesting technically. Called the Cucciolo, it had a four-stroke engine, which gave it a distinct advantage over the two-stroke opposition.

The Cucciolo's valves were operated by pullrods, rather than pushrods, and it had an integral two-speed, pre-selector gearbox and all-metal clutch. Launched in 1945, the engine unit, exhaust, and fuel tank were supplied as a kit for the customer to fit to an existing pedal cycle.

THE U-FRAME MOPED

By the early 1950s, several manufacturers, including NSU, Mobylette, Zündapp, Heinkel, and Ducati, were offering complete machines, using specially-made U-frame structures.

The two most interesting designs were both German: the NSU Quickly and Heinkel Perle. The first Quickly, the N,

appeared in 1953; when production ceased in 1962, almost 540,000 had been manufactured – the total for all variants of the Quickly is an amazing 1.2 million!

Powered by a 49cc, piston-port, two-stroke single, the NSU offered a combination of simple style, quality, and low price. By contrast, the Heinkel Perle (also introduced in 1953) was very advanced, being the first of such small machines to have a cast-alloy frame and swinging-arm rear suspension. A high price restricted sales. Other major U-frame mopeds included the Japanese Honda Graduate and PC50, French Mobylette, Austrian Puch Maxi, and Italian Vespa Ciao.

SPORTS MODELS

The next stage in the evolution of the moped was the sports model. In reality, this was a miniature motorcycle, but with a 50cc engine and pedals. Early examples included the German Victoria Avanti (1959) and Ducati 48 Sport (1963). In the 1970s, high-performance examples appeared, including the Garelli Rekord and Tiger Cross, Yamaha FS1E, Suzuki API, Malaguti Calvacone, and Gilera Touring and Trail. Eventually, legislation was imposed to limit performance, and a more practical kickstarter replaced the awkward pedal starting.

Sales of all mopeds declined as the 1980s dawned, their place being taken in the 1990s by a new breed of 50cc automatic scooter.

below Ducati's Cucciolo was a clip-on engine kit for motorizing a pedal cycle. The name means "little pup" in Italian.

facing page The German Victoria Avanti was designed toward the end of the 1950s. It was an early attempt at providing a sports moped, a genre that would peak in popularity during the 1970s.

above From the early 1960s, the Japanese Honda company built mopeds for the European market at a new production facility in Belgium. Typical was the Graduate; this is a 1973 model.

below The Malaguti 49cc Calvacone, an enduro-styled ultra-lightweight motorcycle, was built in the mid-1970s. It was offered with pedals in several countries, to qualify as a moped.

EDWARD TURNER

Born in London on 24 January 1901, a few hours after the death of Queen Victoria, Edward Turner was given his Christian name because he was the first true Edwardian of seven children – he had three sisters and three brothers. His father owned a light-engineering business, so with this background it comes as no surprise to learn that Edward set up his own motorcycle dealership in south London. Subsequently, he was head-hunted during the late 1920s by Charles Sangster, then boss of the Birmingham-based Ariel company.

A MOVE TO TRIUMPH

At Ariel, Turner designed the legendary Square Four. During the early 1930s, Jack Sangster succeeded his father and subsequently acquired the Triumph marque. He moved Edward Turner to his new acquisition, installing him as managing director and chief designer. Turner's first creation for Triumph

was the range of Tiger singles (250, 350, and 500cc), which amply demonstrated his outstanding styling prowess. Quite simply, he had taken the existing Triumph overhead-valve singles, which had been designed by Valentine Page, and equipped them with tuned engines and additional brightwork. The latter included chromed tank panels, while the remaining body parts were finished with silver paintwork. The transformation turned them into best-sellers.

In 1937, Turner's most famous creation, the Speed Twin, was introduced. It was followed a year later by its more highly tuned brother, the Tiger 100. These two models, equipped with the same 500cc, parallel-twin engine, really cemented his reputation within the industry.

During World War II, the original Triumph factory in Coventry was destroyed by bombing, but a new facility was built a few miles away, in the village of Meriden.

facing page, top left The headlamp nacelle was a Turner styling feature.

left For 1963, Turner and his team revamped the long-running Triumph vertical twin, creating the new unit-construction range of 649cc models, including the 6T Thunderbird.

above A Triumph brochure from 1951, showing the 6T Thunderbird (foreground) and 5T Speed Twin.

above right A 1960 Triumph T120 Bonneville. This model, with the pre-unit engine, was an icon of the 1960s biking scene.

right Edward Turner (right) with Bill Johnson (American West Coast Triumph importer), outside Triumph's factory at Meriden, July 1948.

THE THUNDERBIRD

In late September 1949, Turner's next significant design made its debut, the 649cc (71.0x82.0mm bore and stroke) Thunderbird. It was launched in a blaze of publicity, the first three production models being taken to the Montlhéry circuit in France, where they covered 800km (500 miles), finishing with a flying lap at over 160km/h (100mph).

A brand-new Triumph single appeared in November 1952. This was the 149cc (57.0x58.5mm bore and stroke), overhead-valve, unit-construction Terrier. From the Terrier came the 199cc (63.0x64.0mm bore and stroke) Tiger Cub, which was introduced in 1953 and remained in production until the late 1960s.

The twin-cylinder range received the Tiger 110 (the sports version of the Thunderbird) for 1954, followed by the even quicker T120 Bonneville for 1959. The 3TA Twenty One was

the first unit-construction Triumph twin, being launched in 1957; eventually, all Triumph twins followed this route.

Turner did have the occasional failure, however, such as his sprung-hub rear suspension, which appeared in the immediate postwar period.

Turner's other great gift was his ability to manage costs – his designs made money, which ensured that his time as the company's boss (postwar until the mid-1960s) was Triumph's most successful period. He died on 15 August 1973.

REAR SUSPENSION

It is generally agreed that swinging-arm rear suspension was one of the most significant technical advances in motorcycle design. One of the very first manufacturers to use such an arrangement was the London-based Matchless concern in 1919, on its Model H V-twin. Then, in 1936, Velocette employed a pair of Dowty Oleomatic air rear suspension units on its grand prix racing machines, but it was the famous Ulster engineer Rex McCandless who was responsible for producing the definitive modern version, which worked with a pair of hydraulically operated shock absorbers. Developed in Belfast during World War II, the suspension system was originally referred to as the "spring heel".

INCREASED CONTROL AND COMFORT

Prior to the arrival of swinging-arm, twin-shock rear suspension, riders had been forced to put up with rigid frames with sprung saddles, very basic "boxed" springs (as used by the likes of Gilera and Moto Guzzi), or plungers. The last had been pioneered by the works Norton racers (1936) and on the immediate prewar Ariel Square Four.

These suspension types were usually completely undamped and they offered very little improvement over the rigid set-up. The McCandless swinging arm, however, with its pair of hydraulically damped shock absorbers, gave the rider a much greater level of comfort and superior control of the machine.

right The sprung-hub was not one of Edward Turner's better ideas. Introduced in 1946, it proved a dead end where rear springing was concerned.

below right Much more successful was swinging-arm rear suspension. This arrangement was adopted by Royal Enfield after being tested on a trio of trials bikes during early 1948.

THE PRINCIPLE OF LEVERAGE

The swinging-arm system worked on the principle of leverage. At the forward end of the arm, which was attached to the frame just behind the engine and gearbox, was a pivot. Usually, this was a strengthened steel pin that passed from one side of the machine to the other through a pair of bushes.

The bushes were normally manufactured from phosphor-bronze, but sometimes fibre or plastic was employed. Silent-bloc bushes (a combination of rubber and metal) were also tried, but these had the disadvantage of allowing the pin and swinging arm to flex slightly, which was detrimental to handling and road holding.

SINGLE SHOCKS

Another suspension system was the cantilever type, which was pioneered by the Vincent marque. It was updated by Yamaha during the 1970s.

From the early 1980s, the conventional McCandless-inspired, twin-shock, swinging-arm arrangement was gradually replaced by the monoshock variety, with only one shock absorber. At first, these suspension systems tended to be based on the Yamaha layout, but eventually the shock absorber was mounted vertically. This set-up became known as the rising-rate system. Usually, it provides both pre-load and rebound adjustment.

left Cantilever monoshock rear suspension on a 1976 Yamaha TZ750 racing bike.

below left BMW's patented Monolever single-shock rear suspension system, seen here on a 1990s GS1100.

GERMAN DESIGN

Following World War II and the division of Germany, there were distinct differences in the motorcycle designs that came from companies in the West and East.

THE WEST

In West Germany, civilian motorcycle production did not resume seriously until 1947. At that time, the standard of living of the majority of the population was barely above subsistence level. Not surprisingly, the early postwar machines tended to be simple utility models. In the early 1950s, however, the accent switched to technical innovation, performance, and luxury features, all of which reflected the massive improvement in living standards. By the mid-1950s, West Germany's economic revival was in full swing. The main West German motorcycle marques of the era were Adler, BMW, DKW, Hercules, Horex, NSU, TWN, Victoria, and Zündapp.

Like many West German manufacturers, Adler specialized in two-stroke engines, although the bikes had unusual transmission systems. The 200 and 250cc twins featured an engine-speed clutch mounted on the crankshaft, outside the primary drive, which was by helical gears. The 98 and 125cc singles employed an even more unusual layout, however, the engine being offset to the right, and the clutch and gearbox to the left. Drive to the rear wheel was taken from a sprocket adjacent to the clutch, between the engine and gearbox and concentric with the crankshaft. This was unique to Adler.

NSU offered the Max and Super Max. Displacing 247cc, these overhead-camshaft singles featured a type of valve gear that was unique among motorcycles. Designed by Albert Roder and called Ultramax, the overhead valve gear was driven by long connecting rods, which were housed in a tunnel cast integrally on the left side of the cylinder barrel.

Of all the West German marques, however, Victoria was the most innovative. It produced not only the 347cc Bergmeister narrow-angle, transverse V-twin (introduced in 1951), but also the 197cc Swing. Hailed at its 1956 launch as revolutionary, the Swing was the first motorcycle to have push-button gear changing. It also featured a unique form of

BDG 250 L
with double-piston engine

facing page Launched in 1934, the Zündapp K800 had a flat-four engine displacing 804 cc. Weighing in at 212 kg (467 lb), it could reach 125 km/h (78 mph).

left The German Triumph marque (known as TWN – Triumph Werke Nurnburg – in many countries) produced a series of excellent two-stroke models during the 1950s, including the BDG 250 (shown) and Boss 350 split-singles.

below The BMW 500 cc R50 and 600 cc R69 models were launched at the Brussels Salon in January 1955. Both featured Earles forks and swinging-arm rear suspension.

rear wheel and engine suspension, the engine assembly being pivoted from the frame below the crankcase.

THE EAST

By the late 1950s, the West German motorcycle industry was in decline, its customer base having melted away to purchase cars. In the communist-controlled East, however, there were no such problems. State-owned MZ and Simson enjoyed stability for many decades until the Berlin Wall came down in 1989. At that point, they suffered the same fate as the West German motorcycle industry some three decades before.

Prior to that, both marques had built millions of small-capacity, piston-port-induction, two-stroke, single-cylinder models. Although basic in overall concept, these had been in high demand throughout the old Soviet Empire. In stark contrast, MZ had also produced technically advanced racers, designed by the brilliant engineer Walter Kaaden.

SPANISH STARS

No European country changed so dramatically during the second half of the 20th century as Spain. From a rural, pre-industrial society ruled by a Fascist dictator, Spain became a modern, thrusting democracy and an integral part of the European Community.

THE BEGINNING OF AN INDUSTRY

The Spanish Civil War (1936–39), which had been won by General Franco's Nationalists, left Spain a divided nation. Unlike most European nations, however, it managed to escape involvement in World War II. And it was during this period that the Spanish took their first hesitant steps toward moderniza-tion – and the creation of a domestic motorcycle industry.

The first two companies, Montesa and Sanglas, had already laid the foundations prior to the end of the war in Europe (spring 1945), giving them a head start over the inter-national competition. General Franco's government did the rest by imposing stringent import bans on foreign vehicles and associated components.

MONTESA AND SANGLAS

By the end of the 1940s, Montesa (founded by Pedro Permanyer and Francisco Bulto) was building a range of single-cylinder, two-stroke machines and competing in grand prix racing. Sanglas, meanwhile, was producing 350 and 500cc, overhead-valve, unit-construction engines. The company also became an early user of the hydraulically damped, telescopic front fork.

EXPANSION

It was during the 1950s, however, that expansion of the industry really took off. This was prompted by the arrival of Derbi, Ossa, and a host of other minor marques, their machines often being powered by Hispano-Villiers engines (British Villiers designs built under licence in Spain).

By the end of the decade, the major players had been joined by Bultaco and Mototrans, the latter producing Italian Ducatis under licence.

FRANCISCO BULTO

The most well-known of all Spanish designers was Francisco Bulto, co-founder of Montesa, who left the company in the spring of 1958 to form his own brand, Bultaco. In time, this became Spain's most influential and most sporting marque.

Bultaco's first design, a 125cc, two-stroke street bike led to an excellent series of roadsters, the TSS range of racers (built in 125, 196, 250, and 350cc guises) and, most significantly, a range of world-beating motocross, trials, and enduro bikes. The Sherpa trials bike, for example, developed with the assistance of legendary rider Sammy Miller, changed the face of one-day events when it appeared in the mid-1960s. Overnight, the previously dominant, big four-stroke singles became obsolete. Moreover, there is absolutely no doubt that Bultaco's success encouraged its rivals, Montesa and Ossa, to follow suit, thus creating a new "Spanish Armada" which, this time, was successful.

facing page, left In 1946, Simeon Rabasa Singla's Derbi company built its first motor-cycle; in 1969, the marque won the 50cc world championship, the first of many successes.

facing page, right Manuel Giro and, later, his son, Eduardo, were responsible for the successes gained by Ossa. This Phantom 250 motocrosser dates from the 1970s.

above One of Bultaco's best-known bikes was the Sherpa trials machine, which appeared in 1965.

right A mid-1960s Montesa brochure illustration, showing (left to right) the Sport 250, Impala 175 and 60cc Micro.

SPORTING LIGHTWEIGHTS

Before the arrival of the Japanese in the world motorcycle market during the 1960s, European manufacturers largely led the industry. This was certainly the case when it came to features such as mechanical innovation, design practices, and variety of engine types (both four-stroke and two-stroke). And without doubt, the Italians were the style kings.

THE ITALIANS MAKE THE RUNNING

As with scooters, Italy made the running in the field of sporting lightweight motorcycles. There were various reasons for this. One was that the Italians had been able to resume production more rapidly than anyone else after World War II because, of the major marques, only Benelli's factory at Pesaro on the Adriatic coast had been extensively damaged during the conflict. Several brand-new designs appeared at the first postwar Milan motorcycle show, in December 1946.

Another reason for Italian dominance of the lightweight market was the nation's love of speed events. While the basic requirement was for everyday transport, a model with a racing pedigree, stylish looks, and a bit of excitement was, in the

Italian designer's mind, likely to sell in larger numbers than a dull looker, no matter how well engineered, comfortable, and practical. The latter priorities served manufacturers well in Great Britain, Germany, and Scandinavia, but not red-blooded Italy. Long-distance road events, such as the Giro d'Italia (Tour of Italy) and Milano-Taranto, plus the Italian Formula 2 and 3 racing series, also proved breeding grounds for high-performance, lightweight sports bikes.

THE RIVALS

Immediately after World War II, the established Italian marques, such as Bianchi, Moto Guzzi, and Gilera, had been joined by a host of newcomers. Some of these were large organizations that had built aircraft previously, but had been prohibited from doing so by the Allies following the conflict – Aermacchi (Macchi), Capriolo (Caproni), MV Agusta (Agusta), and Vespa (Piaggio). While the last of these concentrated on scooters, the other companies produced motorcycles, as did another major military supplier, Ducati, which had built radios and electrical equipment.

below Another notable Italian lightweight was the Aermacchi overhead-valve single with horizontal cylinder. This 175cc Ala Rossa dates from 1961.

bottom Compared with Italian lightweights, British machines looked much heavier. This is a 250cc 1965 Matchless G2 CSR.

facing page Besides its legendary overhead-camshaft singles, Ducati built a family of small-capacity, pushrod singles, including this 1957 98cc Sport. Note the unusual oil cooler ahead of the crankcase.

above Carlo Guzzi's last design, the 1956 Lodola, a 175cc, overhead-camshaft single. Of particular interest was its inclined cylinder and heel/toe gearchange with indicator.

INNOVATIVE DESIGN

Although the established firms built some good machines – for example, the Bianchi Tonale (175cc), Moto Guzzi Lodola (175 and 235cc), and Gilera 125/150cc models – the newcomers really made the running with their offerings. These benefited not only from a "clean sheet" when it came to design and styling, but also from the wealth of engineering talent that was at the disposal of the companies. The results were some true classics, among them the Aermacchi Ala Rossa (175cc) and Ala Verde (250cc); the MV Agusta Disco Volante (Flying Saucer); and the Ducati 98 Sport, 175 Sport, and 200 Elite. Technically, however, Capriolo was probably the most interesting marque, with its three main designs. The first of these was a 150cc flat-twin (BMW style); the second a 75cc, face-cam design, in which the crankshaft ran in line with the pressed-steel frame; the third retained the face-cam layout, but with a more conventional bottom end.

SIDECARS

Almost as soon as the motorcycle was accepted as a serious means of transportation – rather than simply an experimental "toy" – riders wanted to carry passengers. This was achieved by attaching a sidecar.

The Oakley Motor Company is said to have invented the sidecar as early as 1900; within three to four years, development was being undertaken by several concerns. There was some debate initially as to whether a rigidly or flexibly mounted sidecar was the most practical, but eventually the former became almost universal practice. At the time, most sidecar bodies were made of wicker to minimize weight, being little more than an upholstered seat with a protective apron for the passenger.

FAMILY TRANSPORT

Following World War I, there was a strong demand for motorized transport for the family. As a result, many specialist sidecar manufacturers came into being, while many motorcycle producers began offering their own "chairs". In Britain, these included BSA, Matchless, Royal Enfield, Sunbeam, Douglas, Panther, Ariel, Chater-Lea, and Triumph. Of the many Continental European sidecar manufacturers, the two most important marques were German: BMW and Steib.

During the 1920s, many families acquired a sidecar outfit simply to save money, and sidecar design evolved. Whereas the first models were nearly all single-seaters, child-and-adult and finally full double-adult versions followed.

For weather protection, sidecars were fitted with simple windscreens at first, but folding hoods and finally hard tops soon followed. Steel or aluminium was employed subsequently for the bodywork.

The sidecar was also used during this period as an alternative to a light van, manufacturers offering models designed specifically for various trades. The sidecar outfit began to be used in sporting events as well, not only for the likes of hill climbs and trials, but also grand prix racing.

By 1936, there were half a million motorcycles on British roads, of which no fewer than 25 percent hauled sidecars.

THE SIDECAR'S HEYDAY

The sidecar's real heyday, however, was in the immediate postwar period of the late 1940s and throughout the 1950s. At the end of World War II in 1945, established sidecar manufacturers returned to the business with renewed energy. Increased demand also led to many new companies being formed, among them Busmar, Blacknell, and Garrard.

In the mid-1950s, the new plastic material, glass-fibre, began to be employed. One of the first sidecars to enter production using this material was the Watsonian Bambini, a single-seater for lightweight motorcycles and scooters.

In the early 1960s, however, sidecar sales began a slide from which they never recovered. There were two main reasons: the prevalence of sporting motorcycles, many of which were not suitable for a third wheel, and the advent of the affordable small car, such as the legendary Mini.

facing page, left Watsonian was the biggest and most well-known of all British sidecar manufacturers. The company's 1954 catalogue detailed touring, racing, and trials models.

facing page, right A familiar sight on the roads of Europe during the 1950s – a motorcycle and double-adult sidecar. In this case, a 1958 Panther 100 (600cc) and San Remo "chair".

above The 1960 World Sidecar Champions Helmut Fath (driver) and Alfred Wohlgemuth, racing in the Isle of Man TT with their BMW Rennsport 500cc outfit.

right Also pictured in 1960, a Royal Enfield 692cc Constellation and Watsonian New Monark sports sidecar.

VINCENT HRD

The Vincent HRD V-twin has gained a special place in motor-cycle history, thanks to such famous models as the Rapide, Black Shadow, and Black Lightning. Philip Conrad Vincent, known simply as PCV, acquired the HRD motorcycle brand from its founder, Howard Raymond Davies, in 1927.

THE INFLUENCE OF PHIL IRVING

In 1932, PCV hired the Australian Phil Irving as chief designer, the pair famously designing the first brand-new model (a 500cc, overhead-valve single) in a hectic 11-week period. This made its debut at London's Olympia motorcycle show in 1934. The legendary motorcycles that followed are generally agreed to have been the fruits of both men's labour and design skills.

It was Irving, however, who came up with the idea of a V-twin. This had been prompted one day at Vincent's Stevenage works, where he had seen a drawing of the 500cc engine lying on top of another so that the crankcases overlapped.

With funds always in short supply, the concept had distinct possibilities. Although a new crankcase would be needed, existing barrels, heads, con-rods, timing gear, gearbox, and other components could be utilized without major change.

THE SERIES A RAPIDE

The prototype of what would become the Series A Rapide (nicknamed the "plumber's nightmare" because of its plethora of fuel and oil pipes) made its public debut in October 1936. It sported a 47.5-degree, V-twin engine with short pushrods operating overhead valves. Displacing 998cc (84.0 x 90.0mm bore and stroke), this was claimed to produce 45bhp at 5,500rpm. Only 78 examples were built before production ceased in July 1939 as Britain prepared for Hitler's war.

THE SERIES B RAPIDE

Compared with the limited-production prewar model, the post-war Series B Rapide was a much better bike. The engine was considerably different, having a wider 50-degree angle between the cylinders, even though the same bore and stroke dimensions were used. The valve gear was enclosed, while a

facing page Created by Philip Vincent and Phil Irving, the original, Series A 998 cc Vincent HRD employed a 47.5-degree V-twin engine. Postwar, as on this 1950s Series C Rapide, the V angle was increased to 50 degrees.

left A stately setting for a stately motorcycle: a 1950 499 cc Vincent Comet single. Essentially, the machine's engine was half a V-twin.

change had been made from hairpin to coil valve springs. In addition, the prewar design had suffered from gearbox and clutch problems, so these had been replaced by new items.

Another major alteration was that the top tube of the frame had been axed in favour of a one-piece, 16-gauge steel oil tank incorporated into a backbone that ran from the steering head to the dualseat.

THE BLACK SHADOW AND SERIES C

The prototype of the Black Shadow sports model appeared in early 1948, Vincent claiming that it was capable of 201 km/h (125 mph). Later that year, the Series C was introduced in both Rapide and Black Shadow forms.

A major difference between the B and C series was that the former employed the prewar Brampton girder front fork, whereas the "C" had the new Girdraulic fork, which was claimed to be superior to even the telescopic type. The fork blades were manufactured from heat-treated forgings of L40 alloy by the Bristol Aircraft Company.

1949 VINCENT HRD SERIES C BLACK SHADOW SPECIFICATIONS

Engine Air-cooled, overhead-valve, 50-degree, V-twin

Displacement 998 cc

Bore and stroke 84.0 x 90.0 mm

Ignition Magneto

Gearbox Four-speed, foot-change

Final drive Chain

Frame All-steel, integral oil tank

Dry weight 208 kg (454 lb)

Maximum power 55 bhp at 5,700 rpm

Top speed 201 km/h (125 mph)

COMMUTER BIKES

Today, the role of personal transport has been neatly filled by the small car, and it may be difficult to believe that much of Western Europe went to work and play on myriad small-capacity, mainly two-stroke-engined, motorcycles during the period immediately after World War II. That, however, is exactly what took place.

Britain, Germany, France, Spain, and Italy all produced suitable designs. Often, these small machines were powered by bought-in engines, from such manufacturers as Villiers (British), Ilo and Sachs (German), and Minerelli and Franco Morini (Italian). Villiers engines were also manufactured under licence in other countries, including Spain.

BRITAIN

Among the most popular British commuter bikes were the various BSA Bantam models (125, 150, and 175cc). The Bantam first appeared in 1948 as the D1, its 123cc engine being an obvious copy of the German DKW RT125; the last B175 model left the production line in March 1971.

The origins of the Royal Enfield two-stroke series could also be traced back to the DKW. There was a wide range of models, stretching from the wartime 126cc ML to the 148cc Prince, production of which finally ceased in 1962.

Although the BSA and Royal Enfield models had employed "in-house" engines, the vast majority of British commuter bikes used Villiers engines. The most important of these machines came from Ambassador, Excelsior, Francis-Barnett, Greeves, James, Norman, and Sun.

GERMANY

Like Britain, the West German motorcycle industry – as opposed to that of the communist-controlled Eastern zone – had a large number of producers building commuter bikes. These included Adler, Ardie, DKW, Dürkopp, Express, Hercules, Maico, NSU, TWN (Triumph), and Zündapp. Many of these marques were industry pioneers.

Technically innovative, although not a sales success, was the unusual Imme (1948–51). Designed by Norbert Riedl, it featured a 98cc, two-stroke engine with a single horizontal cylinder. The engine assembly and rear wheel were suspended as a unit, pivoting below the seat, while the connecting frame also formed the exhaust system.

ITALY

Italian manufacturers built a large number of small-capacity, two-stroke motorcycles during the late 1940s and throughout the 1950s, together with massive quantities of scooters and mopeds. They came from the likes of Aermacchi, Benelli, Bianchi, Guazzoni, Morini, Moto Guzzi, and Parilla.

Unusually, given its well-known, classic four-stroke models, Guzzi produced the most successful of the commuter bikes. It began with the 65cc Guzzino, also known as the Motoleggera. This tiny machine appeared in 1946. It had the traditional Guzzi horizontal cylinder, but instead of the usual piston porting of most small, mass-produced two-strokes, it used a rotary valve, as did the Cardellino (65 and 73cc) and Zigolo (98 and 110cc), which were developed from the same basic design.

left The BSA Bantam was one of the most popular of all commuter bikes, being built in a variety of engine sizes from 1948 to 1971. This is one of the original 125cc D1 models.

facing page, top A typical Italian commuter bike, the MV Agusta 125TEL (Turismo E Lusso), which was built between 1949 and 1954. The final version, seen here, sported a duplex frame, telescopic front fork and swinging-arm rear suspension. The 123cc, two-stroke engine produced 5bhp at 4,800rpm.

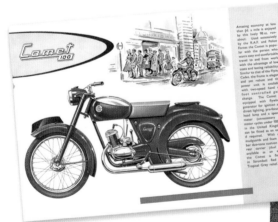

above Built in Greet, Birmingham, the James Comet was one of the least expensive commuter bikes. Powered by a Villiers 98 cc engine with a two-speed gearbox, it offered budget-priced motorcycling during the late 1950s.

right The 1953 German Express M200 was a high-quality machine. Powered by an Ilo 197 cc, single-cylinder, two-stroke engine, it featured a tank-top toolbox, individual sprung saddles and plunger rear suspension.

ENCLOSURE

During the late 1950s and early 1960s, enclosure – concealing parts of a motorcycle behind decorative bodywork, as opposed to adding streamlining to improve performance – gained favour with many manufacturers.

MILLER-BALSAMO

Well before the fashion for enclosure took hold, the Italian Miller-Balsamo concern produced two such motorcycles. The first (a 200cc two-stroke) appeared just before World War II and had panelling around the engine, and one of the star exhibits of the 1946 Milan show was the Jupiter. This revolutionary model carried the practice further with full enclosure of engine and frame. Powered by a 246cc, overhead-valve, single-cylinder engine, it was also notable because of its hydraulically controlled rear suspension and pneumatic front fork. In addition, it had a gear indicator on the instrument console.

NORTON SILVERFISH

Next came the Norton Silverfish. Also known as the "Kneeler", this, like the Featherbed frame, was the work of Ulsterman Rex McCandless. The Silverfish used enclosure to the full, together with streamlining. The alloy bodywork featured a nose-piece that extended from a point about hub level ahead of the front wheel to (at the top) just above the handlebars, in line with the steering axis, and (at the bottom) to the cylinder-base level. The nose was wide enough for the front wheel to turn within it. Behind the nose, separate shields extended rearward, being profiled to the rider's thighs and arms. The rider adopted a semi-prone position – his shins were horizontal and his feet braced against the footrests adjacent to the rear-wheel spindle. In other words, the rider knelt on the bike. The rear was enclosed in a streamlined tail-piece.

Although not a success in racing, the Silverfish did break the one-hour world speed record in November 1953, with an average speed of 215 km/h (133.7 mph).

AERMACCHI CHIMERA

Star of the 1956 Milan show, the Aermacchi 175 Chimera had evolved from a sketch of "the ultimate motorcycle" drawn by Count Mario Ravelli, a notable automobile stylist of the time.

The most advan...
in motor cycle...

Most attractive fr...
instruments, and ...
Two-way parking lamps
Neutral indicator lamps
cables are completely enc...
spacious locker for parcels...

The Petrol Tank Filler Cap, B...
and Tool Kit are neatly arr...
Dualseat, which can be lock...
position.

The metal panniers and rear carrier blend perfectly
with the general styling of the machine and
provide ample luggage space. The plastic holdalls
can be pre-packed, fitted into the Pannier, the
lid replaced and locked. Front and rear Flashers
with handlebar switch can be fitted as an aid to
safe driving.

ARIEL MOTORS LTD.
BIRMINGHAM 29
ENGLAND

facing page Norton works rider Ray Amm with the innovative Silverfish. Although not a success as a grand prix racer, its enclosed and streamlined bodywork allowed it to set a new one-hour world speed record in November 1953.

left An Ariel Leader brochure from the late 1950s, showing details of the machine's enclosed bodywork.

above Several firms offered rear enclosure on their machines. Triumph's design was commonly known as the "bathtub". This is a 1961 650cc Thunderbird.

below Royal Enfield's effort was the Airflow, which offered a combination of streamlining and rider protection.

The details were filled in by Alfredo Bianchi, Aermacchi's design chief. The Chimera had die-cast aluminium panels enclosing the engine and frame, and inspired similar styling in other machines.

BRITISH EFFORTS

There is no doubt that the British efforts of the late 1950s and early 1960s were strongly influenced by the Chimera. The first of these, the Ariel Leader, took the concept a stage farther, however, by providing complete rider protection and also pioneered the use of plastic components on motorcycles. Designed by Valentine Page, the Leader was a 250cc, two-stroke twin.

Triumph and Norton offered rear enclosure for their twins, while Velocette enclosed the bottom half of its 350 and 500cc singles. Royal Enfield favoured the streamlined approach for its Airflow models, but without the rear enclosure.

Subsequently, enclosure went out of fashion, and then made a return in the mid-1980s with the Ducati Paso and the Honda CBR600/1000 models.

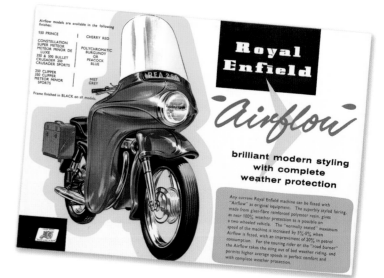

Airflow models are available in the following finishes:

150 PRINCE | CHERRY RED
CONSTELLATION
SUPER METEOR
METEOR MINOR DE LUXE | POLYCHROMATIC BURGUNDY OR PEACOCK BLUE
350 & 500 BULLET
CRUSADER 250
CRUSADER SPORTS
250 CLIPPER
350 CLIPPER
METEOR MINOR SPORTS | MIST GREY

Frame finished in BLACK on all models.

Royal Enfield *Airflow*

brilliant modern styling with complete weather protection

Any current Royal Enfield machine can be fitted with "Airflow" as original equipment. The superbly styled fairing, made from glass-fibre reinforced polyester resin, gives as near 100% weather protection as is possible on a two wheeled vehicle. The "normally seated" maximum speed of the machine is increased by 5%-9% when Airflow is fitted, with an improvement of 20% in petrol consumption. For the touring rider or the "road burner" the Airflow takes the sting out of bad weather riding, and permits higher average speeds in perfect comfort and with complete weather protection.

THE CZECH INDUSTRY

With France, Belgium, and Germany, Czechoslovakia was one of the original pioneering nations of the motorcycle. The famous Laurin & Klement marque, founded while the country was still part of the Austro-Hungarian Empire, was a serious threat to the likes of the Paris-based Werner brothers and the Belgian Minerva company at the turn of the 20th century. In 1918, after World War I, Czechoslovakia became an independent republic.

The two most famous Czech motorcycle brands are CZ and Jawa, although there have been many other marques over the years – almost a hundred of them.

CZ

Ceska Zbroiouka (CZ) was formed in 1918 to manufacture armaments and didn't begin building motorcycles until 1932.

Although the company produced street bikes and road racers, its most famous series after World War II was a family of motocross machines. All were piston-port, single-cylinder, two-stroke models, with twin exhaust pipes at first, then with a single high pipe.

The twin-pipe bike made its first appearance at the 250cc World Motocross Championships in 1960, ridden by Vlastimil Valek, who finished ninth. In 1964, CZ began selling the twin-pipe to private owners under the code 968, followed a year later by the very similar 360cc 969 version. At first glance, the larger-engined machine appeared simply to be a bored-out 250, but this was not the case. The smaller engine had a bore and stroke of 70.0x64.0mm; the 360, 80.0x72.0mm.

In 1967, CZ introduced the single-high-pipe models in world motocross competition. The 968 was updated to

left Jawa entered speedway with its two-valve 500DT, which was based on an existing ESO design. Later, like the 898 model of the early 1980s, it was given four valves.

right Jawa celebrated 50 years of motorcycle manufacturing in 1979. Its first machine had been based on the German Wanderer.

below During World War II, Jawa engineers managed to design a brand-new 250cc model of advanced design under the noses of the occupying Germans. This emerged, post-war, as the Springer – due to its plunger rear suspension and telescopic front fork.

50 YEARS OF JAWA MAKE 1929 1979

JAWA
MOTOKOV, PRAHA

become the 980, which, like the 360, had an almost-square aluminium cylinder. The revised 360 became the 969.01 model. It won the 1967 500cc championship, repeating the feat in the following year, when it was joined as champion by the 250. Both machines were standard bikes, identical to those sold to private owners. Thus, CZs became known as "out-of-the-crate racers".

JAWA

Jawa was set up in 1929 and employed the British designer George Patchett as chief engineer between 1930 and 1939. During World War II, although the country was under German occupation, the company managed to prepare for the postwar era by designing an ultra-modern, unit-construction, 248cc, two-stroke single, with twin exhaust pipes, automatic clutch,

telescopic front fork, and plunger rear suspension. Known as the Springer, it was joined in 1948 by a 350 twin, having the same cycle parts. At the end of 1953, the Swinger 250 and 350cc models appeared, utilizing the same engines. The main difference was that they had swinging-arm rear suspension. These machines were so good that they continued in production, almost unchanged, until the early 1970s.

Jawa began speedway racing in the 1960s with the 500DT (Dirt Track). Subsequently, the overhead-valve, single-cylinder engine was replaced by an overhead-camshaft unit, originally with two valves, and later four. Other features included total-loss lubrication, dry multi-plate clutch, rigid (unsprung) frame, and telescopic front fork. By the end of the 1960s, Jawa had become the world's largest producer of dirt-track (speedway, long-track, and ice racing) bikes.

CAFÉ RACER CULTURE

The "Swinging Sixties" was a period when, for the first time, the young had money to burn. In Britain, motorcycle manufacturers gave them something to burn it on with the latest batch of high-performance bikes, such as the BSA Gold Star, Norton Dominator, Royal Enfield Constellation, and Triumph Bonneville.

Films such as *The Wild One* and *The Leather Boys*, together with the potent pop music phenomenon rock 'n' roll, led to a new breed of young motorcyclist and a new breed of motorcycle – the café racer – which looked and performed more like a competition machine than a road bike.

RACING INFLUENCE

What really helped set the café racer apart from the rest was the influence of British short-circuit racing, which prompted many riders to attempt to re-create on the street what the racers did on the track. Moreover, some manufacturers built clubman's racers (such as the BSA Gold Star), many of which found their way onto the public highway. These, together with suitably modified touring models, formed the era's "ton-up" fraternity – road bikes capable of 160 km/h (100 mph). Racing components such as lightweight alloy or glass-fibre fuel and oil tanks, clip-on handlebars, rear-set foot controls, and the like became the hallmarks of the café racer.

THE RIDERS

Like their bikes, the riders also stood out, in their leather jackets, jeans, leather flying boots, and "pudding basin" or "space" helmets. Often, they congregated at all-night transport cafés (hence the name) and coffee bars, sipping steaming cups of frothy espresso to the blare of rock 'n' roll from the inevitable brightly lit jukebox. To make life interesting, they devised the practice of death-defying "burn-ups" from one café to the next. Or groups would make high-speed trips out of town, to the coast, or to spectate at race meetings, for example.

A COTTAGE INDUSTRY

A vibrant cottage industry sprang up to supply café racer goodies for those who wanted to convert an existing bike or to build a "special" by combining parts from different machines – notably the legendary Triton (a Triumph engine in a Norton Featherbed frame). Ready-made specials, such as the Dunstall Dominator and Rickman Metisse, were also available.

Only the arrival of large-capacity Japanese motorcycles, spearheaded by the 1969 Honda CB750, finally brought the café racer era to an end. Although many of its young devotees were villified, in retrospect, it seems an innocent time – without the excesses of many modern youngsters.

facing page To many, the BSA Gold Star DBD34 Clubmans was the ultimate café racer. It was, and still is, a motorcycle icon. Even standing still, it looks a fearsome beast.

top The famous Ace Café in North London – typical of the all-night transport cafés of the 1960s and a magnet for young British bikers of the day.

above Royal Enfield was one of the few British motorcycle companies to embrace the cult of the café racer, as this 1965 brochure reveals.

right The five-speed Royal Enfield Continental GT of the mid-1960s. Features included factory fitted clip-ons, flyscreen, rev-counter, racing-style tank, and bright red paint job.

5 THE RISI

NG SUN

above Early Honda designs, such as this 1964 C95 150 cc twin, were notable for their advanced specification, which often included such features as push-button starting.

left A 1956 Hurricane 350 cc single. The cycle parts and styling were modelled on the British Triumph marque.

facing page A young Barry Sheene, seen with Suzuki GB sales director Maurice Knight, taking delivery of his new GT550 triple, circa 1974.

THE RISING SUN

Between 1945 and 1959, the Japanese created a vast motorcycle industry, which, as the 1960s dawned, was ready to take on the world. Led by Honda, the Japanese manufacturers made their international debut, more or less simultaneously, on the racing circuit and in the showroom.

When they first appeared in Europe and North America, Japanese racing bikes were only moderately successful. Similarly, the production models, although technically innovative and loaded with standard features that would have been considered extras on native machines, appeared somewhat strange to the Western eye.

Within a few short years, however, this situation had changed drastically. Not only was almost every class of grand prix racing dominated by the Japanese, but also the likes of Honda, Bridgestone, Suzuki, Hodaka, and Yamaha were well-respected names, enjoying the strongest dealer network in the industry, in Europe, and North America.

So what had caused such a major reversal of fortunes? To begin with, the Japanese had created a firm foundation within their domestic market, which, by the end of the 1950s, was one of the strongest and fastest growing in the world. Next, they had planned a properly funded export drive, which included a substantial racing budget. In addition, they had recruited the right people in the various

countries they had targeted for their sales invasion. Never in the entire history of motorcycling had one nation achieved such dramatic inroads into worldwide markets in so little time, and with such determination and efficiency.

But this was no mere marketing exercise. There were sound reasons behind the Japanese success story. Without exception, Japanese motorcycles featured an all-round excellence that neither the European nor American bike builders could match. Their strong performance, good build quality, high level of technical refinement, and value for money combined to make a winning formula. They could only be faulted in roadholding and handling, deficiencies that

were mainly due to "pogo-like" rear suspension units and poor tyre compounds.

Disregarding Honda's C100 Super Cub commuter, the top-selling motorcycle of the first generation of Japanese exports was the 305 cc Honda CB77, known in the USA as the Super Hawk. Even though it was a sportster, it bristled with technical refinement – sophistication even. Included in its comprehensive specification were such features as a push-button starter, a comfortable dualseat, a combined speedometer and rev-counter, twin rear-view mirrors, large silencers, 12-volt electrics, a proper air filtration system, and comprehensive mudguarding.

EARLY DESIGN

Today's Japanese motorcycle is a product of vast manufacturing facilities, features cutting-edge technology, enjoys worldwide sales, and plays a major part in most forms of motorcycle sport. Its almost unchallenged dominance is strictly a recent phenomenon, however, since before World War II very few people outside Japan knew that motorcycles were manufactured there; fewer still had seen, let alone ridden, one.

THE ORIGINS OF AN INDUSTRY

The first machine that could be classified as a motorcycle was imported into Japan from Germany in 1899. It was a steam-powered device with a pair of large wheels shod with solid rubber tyres, and a pair of small outrigger wheels.

In 1908, Narazo Shimazu designed and built the first motorcycle engine in Japan, fitting it into an imported Triumph chassis. The first all-Japanese design was constructed by the Miyata company in 1913. It was powered by a two-stroke engine and sold under the Asahi (Sunrise) name.

By the mid-1920s, vast crowds were watching dirt-track and road racing events in Japan, and, in 1927, Kenza Tada became the first Japanese rider to race in the Isle of Man TT. Most bikes were imported, however, including Tada's British Velocette. This situation continued until the outbreak of war in 1941.

A NEW START

World War II changed everything, leaving Japan a shattered country, with most of its production facilities and cities in ruins. In a few short years, however, an economic and industrial miracle came from this scene of desolation.

First into production were Meguro, Miyata, and Rikuo, closely followed by Tohatsu, Pointer, Abe Star, Hosk, Olympus, and Showa. Some marques had quaint names, such as Pearl, Queen Bee, Jet, Pony, Hope Star, and Happy. Several, among them Hosk and Hurricane, clearly followed British designs, while Fuji and Honda leaned toward German technology.

By 1952, a peak of around 120 marques had been recorded, but few survived until the late 1950s. Those that did were headed by Honda, Suzuki, and Yamaha. Others included Meguro (later taken over by Kawasaki), Tohatsu, Pointer, Bridgestone, Lilac, Liner, and Rabbit.

RAPID GROWTH

Although there were fewer Japanese marques in production after the war, more bikes were manufactured. In 1945, for example, only 127 machines had been built, but by 1950 production had risen substantially to 2,633. Five years later, this figure had increased to 204,304. And by 1960, it had reached an amazing 1,349,000 motorcycles.

By this time, Japanese design had largely moved away from outside influences. This new-found independence coincided with grand prix racing campaigns, employing new multi-cylinder machines, and technical innovations such as push-button starting, direction indicators, improved electrics, and race-bred performance.

facing page The Sunyu 250 cc, overhead-camshaft single was typical of Japanese motorcycles of the late 1950s.

left Built around 1959, this Lilac MF39 had a narrow-angle V-twin engine, modelled very much on the German Victoria Bergmeister. It featured shaft drive.

below left The 1950s Hosk 500 cc, overhead-valve, vertical twin was similar in style to British designs of the period.

bottom left One of the early pioneers of the Japanese industry was Meguro, later taken over by Kawasaki. This is a typical Meguro 500 cc, overhead-valve model of the immediate postwar era.

THE HONDA MOTOR CO.

Soichiro Honda was born on 17 November 1906, the son of a blacksmith, in Komyo, long since swallowed up by the urban sprawl of modern-day Hamamatsu. From an early age, he showed an interest in anything mechanical, and at school displayed a distinct leaning toward practical matters.

HONDA'S EARLY CAREER
After leaving school in 1922, Honda served his apprenticeship as a car mechanic. In the early 1930s, he opened his own garage and did well enough to take up car racing. This ended abruptly, however, after a serious accident in 1936. He sold the business and began manufacturing piston rings.

At the outbreak of World War II, Honda developed a machine so that piston rings could be produced by unskilled women. Toward the end of the war, in early 1945, his factory was heavily bombed, and after the conflict he sold out to Toyota, taking a year off.

THE FIRST MOTORCYCLES
In October 1946, Honda set up the impressive-sounding Honda Technical Research Institute. In fact, it was little more than a wooden hut, on a levelled bomb site on the outskirts of Hamamatsu.

Postwar, the Japanese were in dire need of personal transport, provided it was cheap, and Honda had discovered a cache of 500 small engines (intended for military generators). By adapting these tiny power units and fitting them to conventional pedal cycles, he took his first step toward becoming the world's largest motorcycle manufacturer, which he achieved in little more than 15 years.

Soon he was overwhelmed with orders, and the stock of engines dwindled rapidly. So, to meet demand, he set out to design and build a petrol engine of his own. The result was the 49cc, two-stroke, single-cylinder Type A, which appeared in 1947. When the Honda Motor Company was incorporated in September 1948, the Type A had obtained a 60 percent share of the domestic market.

THE FIRST FOUR-STROKE
The Type E, introduced in mid-1951, was Honda's first four-stroke motorcycle. It featured an unusual three-valve arrangement (two inlet, one exhaust) and two carburettors. Although the newcomer set a new record for Japanese motorcycle production of 130 units per day, Honda was not without problems and, in 1953, he was almost forced out of business due to cash-flow problems, only to be saved by his bank.

Honda was a fighter, however, and a series of new designs helped the company progress throughout the 1950s.

SUPER CUB
The breakthrough came in 1958, with the debut of the C100 Super Cub commuter bike. Introduced in October that year, it sold in vast numbers – production in 1959, for example, reached an incredible 755,589 machines.

Until this time, the world's motorcycle makers had largely concentrated on enthusiast models, which effectively limited sales. The new 49cc (40.0x39.0mm bore and stroke), overhead-valve, "step-through" single gained Honda a vast new market from the man (and woman) in the street. By 1983, a staggering 15 million Super Cubs had been sold.

left A 1964 Honda C95 tourer, with overhead-camshaft, twin-cylinder engine, leading-link front fork, electric starter, full-width alloy hubs, direction indicators, and whitewall tyres.

above Honda's research-and-development department was a vast enterprise that could react to market demands more rapidly than any other motorcycle company during the 1960s.

above right The 49cc "step-through" Super Cub was a runaway success for Honda.

right In addition to its success in grand prix racing and in the showroom, Honda did well in production machine racing. Here, Bill Smith is piloting a CB72 250 in the 1962 Thruxton 500-mile endurance race.

above The 1968 Suzuki AS50 had the marque's Posi-Force lubrication system.

right Among the other advanced features of the Suzuki AS50 were rotary-valve induction and a five-speed gearbox.

LUBRICATION ADVANCES

Although advances in performance were exhibited by many Japanese designs, an important – and often overlooked – feature was the increased sophistication of the product. Nowhere was this more apparent than in two-stroke engine lubrication.

THE BACKGROUND

Four-stroke and two-stroke motorcycle engines of conventional design have entirely different lubrication systems. The former have pressure-fed crankshaft bearings, so inevitably a lot of oil is splashed around the crankcase. Such a system would be impractical in a normal two-stroke, where the crankcase serves as a primary compression chamber for the incoming petrol/air mixture. This would pick up so much excess oil mist that the exhaust would be excessively smoky and the spark plugs would be prone to fouling.

The two-stroke solution is to transfer a small quantity of oil into the crankcase with the mixture. For many decades, two-stroke motorcycle fuel tanks contained "petroil", an oily mixture that was delivered by the carburettor to the crankcase. This primitive arrangement worked surprisingly well, but had the disadvantage of causing a smoky exhaust under light and heavy loads, and reducing lubrication when slowing or shutting the throttle.

THE METERED PUMP

The Japanese, notably Suzuki and Yamaha, pioneered the use of a separate oil tank, which fed a variable metering pump controlled by the throttle, in the modern high-performance two-stroke engine. This load-linked oiling concept was not a Japanese innovation, however – the Italian Adalberto Garelli and the British Velocette marque had already employed the principle. What the Japanese did was to refine the system and make it suitable for mass production.

The basis for the system is a pump driven by the engine, its output being proportional to engine speed. The output is controlled by an adjustable nozzle that is varied in area by a

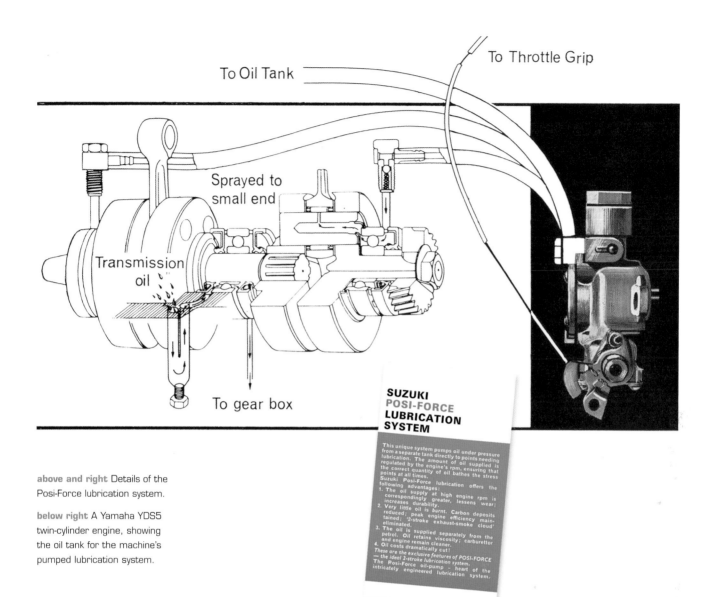

To Oil Tank

To Throttle Grip

Sprayed to small end

Transmission oil

To gear box

above and right Details of the Posi-Force lubrication system.

below right A Yamaha YDS5 twin-cylinder engine, showing the oil tank for the machine's pumped lubrication system.

mechanism linked to the throttle control. This allows the maximum pump delivery to be achieved at full throttle, but progressively restricts output as the throttle is closed. The pump draws oil from the separate tank and delivers it under pressure to the crankshaft bearings and the cylinder walls, from where it reaches the big-ends and other bearing surfaces as a mist or by being thrown against them by centrifugal force.

ENVIRONMENTALLY FRIENDLY

Such an oil-injection system allows oil flow to be kept to the minimum, reducing consumption and pollution, yet allowing the engine's power to be turned on or off without adverse effects. There is no doubt that it has played a major role in ensuring that the two-stroke remains acceptable in the modern, environmentally conscious, world.

GRAND PRIX TECHNOLOGY

During the 1950s, Japanese manufacturers began to show an interest in world championship grand prix racing. Toward the end of the decade, they sent observers to European circuits to study and photograph the latest racing motorcycles.

HONDA STARTS THE BALL ROLLING

Then, in 1959, Honda arrived on the European racing scene with a squad of 125cc, double-overhead-camshaft twins. These were reliable, if not terribly fast, and they won the team prize in the Isle of Man TT.

The following year, Honda returned to Europe with a twin-cylinder 125 and a four-cylinder 250. The former had been modified in several ways since its TT debut, most of the changes affecting the chassis – it had a new frame and a telescopic front fork in place of the leading-link original. A lower centre of gravity and improved streamlining helped further. Although Honda did not win a title in 1960, it did gain ground on the European opposition.

In 1961, a strong Honda team overtook the Italians, the twin-cylinder 125 winning that year and in 1962 – with little modification. Meanwhile, Mike Hailwood took the 250 four to its first world championship in 1961, a feat that was repeated by Jim Redman the following season.

From then on, Honda continued to amaze, with a succession of new, high-tech racers, including a 350 four (1962), a 125 four (1965), a 50 twin (1965), a 125 five (1966), and a 500 four (1966). The opposition was crushed, however, by the 250 six, which appeared in late 1964, followed by a 297 version in 1966.

YAMAHA

Yamaha made its first appearance in European grand prix racing during 1961, with prototype 125 and 250cc models. Clearly derived from the MZ, these "guinea pigs" were being tested prior to a full-scale attempt at the world championships. This occurred in 1963 with the RD56 disc-valve twin. It was not until 1964, however, when Yamaha signed rider Phil Read, that it won the championship.

facing page Yamaha sold twin-cylinder racers that could win at the highest level. Mick Grant is seen here in 1973 with a liquid-cooled example.

right Tohatsu also sold machines with race-winning potential.

below Yamaha's 1966 full works GP RD05A 250cc four, with liquid cooling and eight speeds.

below right During the early 1960s, Honda's four-stroke four dominated the 250cc World Championship.

bottom right Phil Read won the 250cc World Championship for Yamaha in 1964 and 1965.

The arrival of the new Honda 250 six spurred Yamaha to design its 125 and 250cc, 90-degree V4s. After winning the 125 and 250 titles in 1968, Yamaha retired (Honda had quit at the end of 1967).

SUZUKI

Although Suzuki had contested the Isle of Man TT in 1960, it was not until the 1962 season that the marque really began to be competitive. Its increased success at that point was due in no small part to the defection of MZ rider Ernst Degner to the Suzuki camp. Degner and Suzuki became the inaugural 50cc champions in 1962, using the MZ-inspired, single-cylinder, disc-valve, two-stroke single. In addition to the Suzuki 50 (which was later built in twin-cylinder guise with liquid cooling and no fewer than 14 gears) and 125 twins, the company produced the RZ250. This model employed the square-four engine layout – but unlike the famous Ariel design, the Japanese machine was a disc-valve two-stroke with liquid cooling and a six-speed gearbox.

THE KAWASAKI TRIPLE

Of the Japanese "Big Four", Kawasaki was the last to begin manufacturing motorcycles. Unlike Honda, which had been established to build two-wheelers, Kawasaki was a substantial industrial concern already in heavy engineering, making trains, ships, and aircraft.

ENGINES FIRST

During the 1940s and 1950s, Kawasaki had been a major supplier of engines to the fledgling Japanese motorcycle industry. After co-operating with Meguro, it acquired the marque in 1963 and began building its own bikes.

A landmark in the evolution of the all-Kawasaki motorcycle occurred in September 1968 with the launch of the 499cc H1 Mach III (known in Japan as the 500-SS). This was an across-the-frame, three-cylinder two-stroke with piston-port induction (previously, the marque had employed disc valves on its 250 and 350cc twins).

Other features of the newcomer included a pressed-together crankshaft (allowing one-piece con-rods to be used), six main bearings (all ball races) and horizontally split crankcases. Four oil seals were fitted – two at the shaft ends and the remaining pair between the inner main bearings. The three cylinder barrels and heads were separate castings.

UNIT CONSTRUCTION

A five-speed gearbox and wet multi-plate clutch were built in unit with the engine. The lubrication system followed normal Kawasaki two-stroke practice, oil being injected by pump into the inlet tracts from an independent supply. The power output was 60bhp at 7,500rpm, and the machine was endowed with rocket-like acceleration.

A closely related racing version, known as the H1R, was developed by the end of 1969. In addition to a higher state of tune, it had a close-ratio gearbox and a dry clutch. It employed a supplementary "petroil" mixture to improve lubrication.

OTHER VARIANTS

After the original H1 street bike, Kawasaki introduced the H1A (1971), H1B (1972), H1D (1973), H1E (1974), H1F (1975), and, finally, the KH500 in 1976.

While the H1 was exciting, the H2 (Mach IV) was truly awesome. It had the same basic layout, but was equipped with a 748cc engine that produced 74bhp at 6,800rpm. Its performance, at least in a straight line, could not be matched by any other contemporary series-production roadster. This was offset somewhat by extremely high fuel consumption. The 750 H2 was built between 1971 and 1975.

Other Kawasaki triples included the S1 (249cc), S2 (346cc), and S3 (400cc). The S1 model became the KH250 in 1976, the S3 the KH400 in the same year.

The world fuel crisis of 1974 killed the design, however, Kawasaki being forced to explore new avenues in its quest to establish itself in the world market. The result was its series of double-overhead-camshaft, across-the-frame fours, beginning with the 903cc Z1 in 1972.

left Kawasaki works rider Mick Grant with his KR750 triple, circa 1976.

facing page:
top The original H1 (Mach III) arrived in time for the 1969 season and was developed over the following years. The KH500 model (shown) appeared in 1976.

bottom Kawasaki also produced smaller triples, like this KH250 and the KH400.

THE KAWASAKI TRIPLE

YAMAHA LC

In launching its LC (Liquid Cooling) range, Yamaha took several major technological steps forward. Apart from the cooling system, these included cantilever rear suspension, giving the man in the street a new dimension in two-stroke performance, at an affordable price.

The first example of the concept was the RD 350LC, which was launched in October 1979 at the Paris motorcycle show. Designed as a replacement for the long-running RD series, that original 350 was the forerunner of a whole family of machines: RD 250LC, RD 125LC, RD 80LC, 350 YPVS, and, finally, the flagship RD 500LC four.

THE DESIGN

Although some features – including the excellent reed-valve induction system, which provided greater engine flexibility – were shared with the outgoing air-cooled RD range, much was new. Most obvious was the liquid-cooling arrangement. Yamaha used a 50/50 mixture of distilled water and ethylene glycol (anti-freeze) as the coolant, which was forced around the aluminium cylinders by a crankshaft-driven pump. Liquid cooling offered two distinct advantages: it minimized bore distortion – the cylinder had a more uniform temperature – and reduced the amount of mechanical noise.

The cantilever monoshock suspension of the 250/350LC models had been developed from Yamaha's YZ motocrossers. It offered increased rigidity and much longer rear wheel movement compared with the pivoting fork and twin shocks of the previous RD series.

Another major development was the anti-vibration engine mounting system. The motor was pivoted at the rear and retained by high-deflection rubber bushes at the front.

The LC models had many common parts: 350LC barrels fitted the 250LC; the 250 and 350LC barrels fitted the air-cooled RD; and the TZ barrels (up to 250G) fitted the RD and LC. Therefore, it would be logical to assume that the LC engine had been derived from the TZ, but this was not the case. It was a direct development from the RD, bearing little resemblance to the piston-ported TZ racers of the era.

YPVS

The YPVS (Yamaha Power Valve System) arrived on the 350 for 1983, and a year later on the RD 500LC, which had a 499cc (56.4x50.0mm bore and stroke), liquid-cooled, 50-degree V4 engine with two crankshafts. Its reed-valve induction differed from previous Yamaha practice in having two distinct set-ups on the same engine, allowing all four carburettors to be grouped between the banks of cylinders. In addition, the six-speed gearbox differed from the conventional splash-fed variety of the other LCs, having a trochoid oil pump.

above The 1983 350 YPVS (Yamaha Power Valve System). In simple terms, this had a valve that operated in the exhaust-port window, giving an additional 12bhp over the original 350LC.

above The RD 350LC (together with a 250 version) appeared in time for the 1980 season. Aimed at the European market, it set new standards of performance in its class.

left The TZ racer (a G model is shown) pre-dated such details as liquid cooling, a six-speed gearbox, and cantilever monoshock rear suspension.

below left The sensational four-cylinder RD 500LC arrived in 1984. Equipped with a V4 engine, it could reach almost 241 km/h (150 mph).

1980 YAMAHA RD 350LC
SPECIFICATIONS

Engine Liquid-cooled, reed-valve, two-stroke, parallel twin

Displacement 347cc

Bore and stroke 64.0x54.0mm

Ignition Electronic

Gearbox Six-speed, foot-change

Final drive Chain

Frame All-steel, cantilever rear

Dry weight 137kg (302lb)

Maximum power 47bhp at 8,500rpm

Top speed 177km/h (110mph)

6 RACING IMPROVES THE BREED

above Mick Grant winning the 1974 Isle of Man Production TT on the famous Triumph Trident "Slippery Sam". This machine was victorious no fewer than five times between 1971 and 1975.

left The Italian Ducati marque built the limited-production 200 cc Motocross to special order in 1959 and 1960.

facing page Yorkshireman Denis Parkinson and the 499 cc, double-overhead-camshaft Manx Norton, with which he won the 1954 Manx Grand Prix.

RACING IMPROVES THE BREED

In 1949, the Fédération Internationale Motorcycliste (FIM) adopted the structure of an official racing series, leading to the first road racing grand prix world championships. That year, six circuits were used. Only the premier 500 cc solo championship was staged at all six of them, however; the 350 cc championship took place at five; the 250 cc at four; and the 125 cc and sidecars only at three.

In those early days, all the leading contenders employed four-stroke engines, normally with overhead camshafts (single or double) and one or two cylinders. In succeeding years, however, this pattern would change considerably.

The Italian Gilera four-cylinder racer won the 500 cc title in 1950, after the British AJS Porcupine twin had been victorious in 1949, but, surprisingly, a double-overhead-camshaft, single-cylinder Norton won in 1951 and 1952. This was due not only to Geoff Duke's brilliant riding, but also to Norton's excellent frame design. The work of Irishman Rex McCandless and nicknamed the "Featherbed", the frame out-handled anything else on the track.

Duke signed for Gilera for the 1953 season and, although his bike's frame was not an exact copy of the Norton's, the Italian four-cylinder machine benefited from a chassis that clearly showed British influence.

In 1950, the Gilera had been faster than the British bikes, but not until given a decent frame and Duke's unmatched riding skills did it become unbeatable. The

combination took a trio of world 500 cc titles (1953, 1954, and 1955).

In 1956, another Englishman, John Surtees, won on an MV Agusta four, a feat he repeated in 1958, 1959, and 1960. The Italian Libero Liberati won in 1957 on the Gilera. In the smaller classes, the competition was even fiercer, with Italian MVs, Moto Guzzis, Gileras, FB-Mondials, and Ducatis taking on German BMWs, DKWs, and NSUs.

Although the British challenge faded in the grand prix classes, it made a return in Formula 750 (later Superbike), when such motorcycles as the BSA Rocket 3, Triumph Trident, and Norton Commando vied for honours.

The other major motorcycle sport was scrambling (later

known as motocross). During the late 1940s and the 1950s, the premier 500 cc class was dominated by heavyweight four-stroke singles. The 250 cc mounts, however, were usually two-strokes. By the mid-1960s, overbored 250s, such as the 360 cc CZ, Greeves, and Husqvarna, had challenged successfully for honours in the 500 cc division. Only BSA, with its 441 and later 500 cc, overhead-valve, unit-construction model (based on the existing C15 250 cc design), managed to hold off the two-stroke hordes until, finally, CZ became champion in 1967.

Other forms of motorcycle sport included one-day trials, long-distance trials, speedway, grass-track, dirt-track, and ice racing.

THE EUROPEANS

Before World War II, the British and Continental European motorcycle manufacturers – among them Norton, Velocette, AJS, Gilera, Moto Guzzi, BMW, DKW, and NSU – had battled for grand prix honours.

THE WORLD CHAMPIONSHIPS BEGIN

Previously known as the European Championships, the series received world status for the 1949 season, when the titles went to FB-Mondial (125 cc), Moto Guzzi (250 cc), Velocette (350 cc), AJS (500 cc), and Norton (sidecar). As the 1950s began, however, the Europeans began a gradual take-over, Gilera gaining the 500 cc crown. Prewar, Gilera had employed a liquid-cooled, supercharged four, but the postwar bike was air cooled and unblown.

Conceived by Piero Remor (who moved to rival MV in 1950), the Gilera had appeared in 1947, retaining the

across-the-frame configuration of the earlier machine. After signing Geoff Duke for the 1953 season, Gilera dominated the 500 cc championship until it retired at the end of 1957.

FB-MONDIAL

Like Gilera, FB-Mondial had been a title holder until it also quit in 1957. The 125 cc Mondial was the work of Alfonso Drusiani, and the prototype had been ready in mid-1948. The single-cylinder engine's two overhead camshafts were driven by a vertical tower shaft, with straight-cut bevel gears, on the offside of the power unit, while the 80-degree-inclined, special steel valves employed exposed hairpin springs. Both the cylinder and head were cast in light alloy, having an austenitic liner and steel valve seats. Even on the low-grade, 80-octane petrol of the time, the 123.5 cc (53.0 x 56.0 mm bore and stroke), double-overhead-camshaft engine could rev safely to

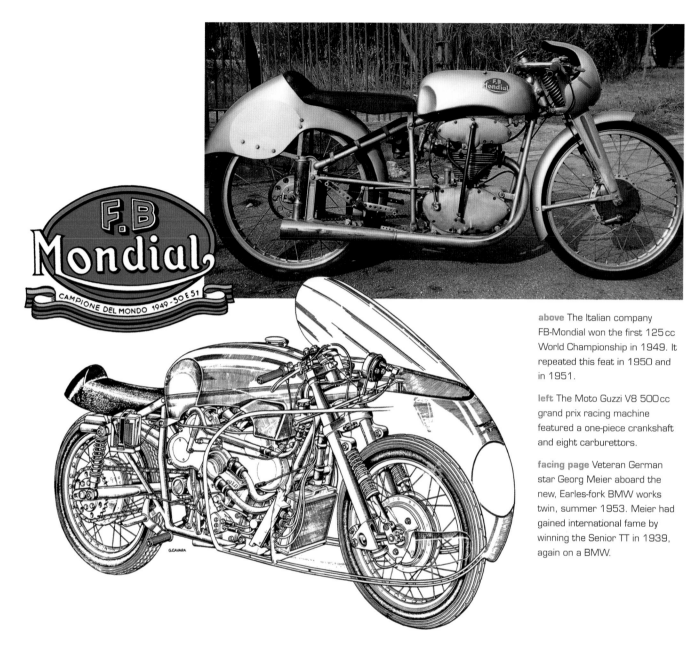

above The Italian company FB-Mondial won the first 125 cc World Championship in 1949. It repeated this feat in 1950 and in 1951.

left The Moto Guzzi V8 500 cc grand prix racing machine featured a one-piece crankshaft and eight carburettors.

facing page Veteran German star Georg Meier aboard the new, Earles-fork BMW works twin, summer 1953. Meier had gained international fame by winning the Senior TT in 1939, again on a BMW.

8,000 rpm and generate 11 bhp, giving the bike a top speed of 150 km/h (93 mph).

Mondial won the 125 cc title in 1949, 1950, and 1951. It also took the 125 and 250 cc championships in 1957. During the same period, first Moto Guzzi, then NSU, and, finally, MV Agusta (except for 1957) gained the 250 cc honours.

DKW'S THREE-CYLINDER CONTENDER

Although it didn't win the championship, the DKW three-cylinder racer was the most technically interesting entry in the 350 cc title stakes. Introduced in 1952 and designed by Erich Wolf, the first version had been created by modifying an existing 250 twin with, at first, a magneto and an inlet disc in the front of the engine. Then Wolf had the idea of fitting a third cylinder in place of the magneto at the front of the crankcase, and the 350 triple was born. As on the 250, the two outer

cylinders were inclined forward by 15 degrees from the vertical to assist cooling at the rear of the cylinders. This resulted in an included cylinder angle of 75 degrees.

BMW RENNSPORT

Unlike DKW, BMW stuck to its traditional transverse, flat-twin engine layout. The definitive Rennsport, however, used by the factory team in 1954, was notable for its fuel injection system (developed by BMW in conjunction with Bosch). Fuel was fed by gravity to a paper cartridge filter mounted on the offside of the crankcase, above the cylinder. From the filter, it passed to a plunger-type pump, which provided a minimum delivery pressure of 40 kg/sq.cm (570 lb/sq.in); there was no direct rider control of the pump delivery.

BMW-powered sidecar outfits won no fewer than 19 world titles in the 21 years, between 1954 and 1974.

left Norton built small batches of its Featherbed-framed Manx model between 1951 and 1962, and examples were raced all around the world. Here, Rudi Allison is shown leading the field in the 1952 South African Natal 100 – racing on dirt.

facing page:

top Rex McCandless' 1949 Norton Featherbed prototype.

bottom Percy May, 1971 British 500cc Champion, the last British champion on a British bike, with the exception of the rotary Norton. It was an amazing feat for what was, essentially, a 20-year-old design.

THE McCANDLESS FEATHERBED

By the end of the 1949 racing season, Norton's singles were facing opposition from Italian four-cylinder machines. To meet the threat and have any real chance of winning future world championships, it was clear to all concerned at the British company that drastic measures would be necessary.

REX McCANDLESS

The solution to Norton's problem was not destined to come from its own technicians, but from Belfast, Northern Ireland, where a young rider and engineer, Rex McCandless, had already created a superior form of motorcycle rear suspension. He had marketed conversion kits to allow this suspension to be fitted to existing rigid-framed machines, but subsequently had sold the business.

Freed from the pressures of running a business, McCandless set himself a new goal – to produce the ultimate racing chassis. A Norton engine was chosen for the project, which is not too surprising given that his great friend, Artie Bell, was a member of the Norton racing team.

McCandless could clearly see, however, that Norton's existing "Garden Gate" plunger frame (dating back to 1936) was well past its sell-by date.

A BRAND-NEW FRAME

Creating a brand-new racing frame was not an easy job and, working in great secrecy, McCandless produced several prototypes. All of them were tested by Bell.

Convinced that he was on the right track, he stuck to the

task. Then came the real breakthrough – he was offered a freelance position with Norton.

Surprisingly, Norton's first commission was not for a road racer, but a trials machine, the 500T overhead-valve single. After a truly awful 1949 season, however, McCandless was asked to design a racing frame for the company.

McCandless' frame was of duplex construction, the main portion comprising two continuous lengths of tubing. Each of these was attached to the base of the large-diameter steering head, from where they curved outward and ran horizontally to the saddle peak, dropped vertically, then ran forward to form the cradle for the engine and gearbox. Finally, they ran up and inward, passing between the top rails to be attached to the top of the steering head. The frame was braced at strategic

points by cross-tubes, while a large gusset on each side provided support for the rear engine plates, footrests, gearchange and brake pedals (the rear brake was cable operated on the racer), and the swinging-arm pivot fork. At the rear, the swinging arm was controlled by a pair of hydraulically damped shock absorbers.

Following track testing, rider Harold Daniell described the frame as a "featherbed", and the nickname stuck.

WORLDWIDE ACCLAIM
Geoff Duke went on to win three world championships with the Featherbed Norton, and the frame received worldwide acclaim in both its "wideline" and subsequent "slimline" forms. It set new standards in handling on the race track and the road.

left The Triumph works trials team of the early 1950s with one of its 500cc Trophy twins. Left to right: Johnny Giles, P.F. Hammond, and Jim Alves.

facing page:

top Günther Baumann rode as a member of MZ's Trophy team in the 1963 and 1964 International Six Days Trials.

bottom During the 1970s, Italian machines from such marques as Fantic, SWM, and Italjet (shown) emerged as serious contenders in one-day trials events.

DIRT-BIKE SPORT

In the early days of the motorcycle, reliability trials were popular, sometimes being as long as 1,609km (1,000 miles). Eventually, they gave way to much shorter one-day events and the famous International Six Days Trial (ISDT), the first of which took place in 1913.

Later, the sport of scrambling (now known as motocross) began in England on 29 March 1924, when the Camberley motorcycle club decided to modify the rules of its popular Southern Scott Trial by deleting the gymkhana section and concentrating on a timed event. Some 80 competitors assembled for the start, and the winner was to be the rider who recorded the fastest time over two heats. Thus, a new sport was born.

MOTOCROSS TAKES OFF

Although both long-distance and one-day trials were popular at both national and international level before World War II, motocross did not become an international sport until after the conflict.

The annual Motocross des Nations was inaugurated in 1947, while, in 1952, the FIM established the European 500cc championship series. In 1957, this became the world championship, 125 and 250cc classes being added later.

THE BIKES

During the 1940s and 1950s, the large-capacity four-stroke ruled supreme, usually with an overhead-valve or overhead-camshaft, single-cylinder engine. British machines included the BSA Gold Star, Matchless G80CS, Ariel HS, Norton Manx, Triumph Trophy, and Velocette. Belgium provided the FN, while Swedish dirt bikes came from Crescent, Lito, Monark, Husqvarna and Lindstrom.

The British Rickman Metisse (from a French word meaning "mongrel bitch") was a revolutionary chassis that proved very popular. By the mid-1960s, many engines – including Triumph, Matchless, and Spanish Bultaco units – had been fitted to it to produce race-winning combinations.

TWO-STROKE TAKE-OVER

As motorcycle technology developed, so did the make-up of the sport. Lightweight, two-stroke bikes made rapid progress from the late 1950s, among them Maico, Monark, Husqvarna,

Jawa, Greeves, and CZ. After the 250cc class was given world status in 1962, the competition became even hotter.

The next advance came in the middle of the decade, when CZ and Greeves built 360cc versions of their lightweights, effectively sounding the death knell for the old heavyweight "thumpers". For a time, BSA responded by developing a larger-engined version of its C15 four-stroke 250, on which Jeff Smith took the 500cc world title in 1964 and 1965. After that, however, it was two-stroke domination all the way.

FRAME AND SUSPENSION

As the power output of both two- and four-stroke engines increased steadily, designers began to pay particular attention to the machines' frames and suspension. Consequently, many different theories on the layout and construction of these major features were conceived and tried. The result was that frames became stronger, yet ever lighter, while suspension performance improved dramatically. The latter was achieved by equipping bikes with longer-travel forks, adjustable damping and, ultimately, monoshock rear suspension instead of the traditional twin-shock arrangement.

TAGLIONI & THE DESMO

For almost 40 years, the Italian engineer and designer Fabio Taglioni was the inspirational force behind the extraordinary success of Ducati motorcycles on the road and the race track. His vision and talent for exploiting often radical ideas produced a series of motorcycles that established the company as the most charismatic of Italian sporting marques.

Born on 20 September 1920, Taglioni joined Ducati in the spring of 1954, following spells at Ceccato and FB-Mondial. His talent as a designer had been demonstrated by the V4 engine he had conceived while a student at Bologna university. It was not until his arrival at Ducati, however, that his talent was exploited to the full.

A DEBUT GP VICTORY

With the victory of Degli Antoni on a 125cc Ducati in the 1956 Swedish Grand Prix, the Bologna-based company became one of the outstanding competitors at the top level of motorcycle racing. Antoni's death in a testing accident at Monza set back Taglioni's racing programme, but momentum

soon returned and, in 1959, a teenage Mike Hailwood won his first grand prix, the 125cc Ulster, on a Ducati.

Both bikes ridden by Antoni and Hailwood had employed desmodromic ("desmo" for short) valve operation. The word "desmo" has long been associated with Ducati and Taglioni. In fact, many wrongly assume that Taglioni invented the system. Its theoretical advantages, however, have been known since the invention of the combustion engine itself. What Taglioni did was to make it work – on the track, and later in series-production roadster engines.

CONTROLLED RUN

"Desmodromic" was coined from the Greek words for "controlled run". The purpose of the design was to eliminate one of the chief problems of conventional sprung valve operation at high engine speeds, the phenomenon of valve float, or "bounce". This was a particular weakness at the time, before the advent of four-valves-per-cylinder technology and the rev limiters found in modern engine management systems.

facing page Englishman Tony Rutter won no fewer than four world TT F2 championships for Ducati during the early 1980s, riding the 600cc Pantah V-twin.

right Mike Hailwood gained many of his early victories aboard motorcycles designed by Fabio Taglioni, including his first grand prix – the 1959 125cc Ulster – on this machine.

below Fabio Taglioni at his drawing-board in the Ducati factory, Bologna.

below right Gianni Degli Antoni with the fully-streamlined, triple-camshaft 125cc Desmo single after winning the 1956 Swedish Grand Prix – the bike's first race.

Valve bounce occurs when conventional valve springs are unable to respond fast enough to close the valves onto their seats. In a desmodromic system, the troublesome springs are replaced with a mechanical closing system, much like that employed to open them, thus providing a positive action. This effectively eliminates the bounce, allowing the engine to rev higher. Some desmo systems employ a separate camshaft (as in the 125cc Ducati Desmo single of the late 1950s) to close the valves, while others may simply have extra lobes on a single camshaft.

At first, the system was used exclusively on Ducati's racers, but subsequently the Italian company adopted it for series production with the introduction of the "wide-case" Mark 3D Desmo single in 1968. By 1980, all Ducati models featured desmodromic valve operation; a situation that continued into the 21st century.

WALTER KAADEN

Generally known as the father of the modern two-stroke, Walter Kaaden was an engineering genius. Born on 1 September 1919, he gained his engineering diploma at the technical academy in Chemnitz. Later, he joined the Peenemunde research establishment, where the Germans developed their V-weapons during World War II.

Although Kaaden fell into American hands at the end of the war in 1945, they handed him over to the Russians. During the early 1950s, he was employed at the IFA factory at Zschopau. This had been the pre-war home of DKW, and from the mid-1950s was occupied by MZ.

DISC-VALVE INDUCTION

Walter Kaaden took charge of the MZ racing department in 1953, controlling all the marque's sporting activities. There, he developed a series of two-stroke engines, all of which had the carburettor mounted on the side of the crankcase, where it supplied the fuel mixture by means of a disc valve set in the crankcase wall. He was also a pioneer of the now widely used 54.0x54.0mm "square" bore and stroke dimensions (on a 125cc single), together with the expansion-chamber exhaust.

At the beginning of the 1953 season, the 125 had produced only 9bhp at 7,800rpm, but by the end of that year it was putting out 12bhp at more than 8,000rpm.

PORTING AND EXHAUST SYSTEMS

Much of the additional power that Kaaden had produced came from extensive changes to the transfer and exhaust ports, the compression ratio, and the exhaust system. In fact, Kaaden had made his most important breakthrough with the exhaust. He had been the first to recognize the importance of a highly resonant exhaust system (chamber). When combined with the extended port timing permitted by disc-valve induction, multiple transfer ports, and a squish-type combustion chamber, this gave huge increases in power compared with conventional piston-port induction.

By 1955, Kaaden had developed the disc-valve, two-stroke single to a point where it was almost on par with the very best of the previously all-conquering, twin-cam four-strokes. His work was largely ignored by the international community, however, as the Cold War was at its height, and East Germany was being treated like a leper.

A TWIN

Kaaden also designed and built a 250 two-stroke twin. His broad strategy was to experiment on the 125, then transfer the technology to the twin. In effect, the 250 was two 125s with their crankshafts spliced to a common primary drive. It was the first MZ to employ rearward-facing exhausts (from the cylinder barrels).

By 1961, the 125 single and 250 twins were offering the equivalent of 200bhp/litre, an amazing figure for the time, and MZ looked set to be world champion. Its top rider, Ernst Degner, defected to the West, however, and went to work for Suzuki. The Japanese manufacturer won its first world title in the following year.

BSA/TRIUMPH TRIPLES

Today, the BSA and Triumph three-cylinder 750s are best remembered for their contribution to the early days of Formula 750 (later Superbike) racing. Anyone who saw the British triples howling their way to victory during the early 1970s is not likely to have forgotten the magical experience.

During the latter half of the 1960s, rumours abounded concerning the introduction of new, larger-engined models from BSA and Triumph. What few observers knew at the time, however, was that the motorcycles in question had been under development for many years.

Finally, in 1968, specific details emerged. Up to that time, both marques had built a series of overhead-valve, parallel twins, at first with separate engines and gearboxes, but from 1962 (BSA) and 1963 (Triumph) with the more compact, unit-construction layout (engine and gearbox combined).

A JOINT VENTURE

Belonging to the same group, the two companies had designed the new bike as a joint venture. Its unit-construction engine, a parallel three-cylinder unit, had a capacity of 740 cc (67.0 x 70.0 mm bore and stroke) and featured pushrod-operated valves (two per cylinder). The gearbox was a four-speed device.

Design work had taken place at the Triumph factory, the bike's chief architects being Doug Hele, Bert Hopwood, and Jack Wicks. Thus, it was not surprising that the engine followed Triumph, rather than BSA, practice. It had been created by adding a third cylinder to the existing (500 cc) twin. The valve gear followed the pattern set by Edward Turner, with two camshafts fore and aft of the one-piece cylinder block.

Driven by gears, these operated pushrods that were enclosed in vertical tubes between the cylinders.

TRIDENT AND ROCKET 3

Upright and inclined (15 degrees) cylinders and different frames were the major distinctions between the Triumph and BSA triples, by now known as the Trident and Rocket 3 respectively. However, they employed the same suspension, brakes, wheel sizes and electrical equipment.

The well-known Triumph factory tester and road racer, Percy Tait, was the first to campaign a racing version of the three-cylinder Triumph (or, for that matter, the BSA), when he took part in an international meeting at Brands Hatch in the summer of 1969.

Both Triumph and BSA competed at Daytona in the spring of 1970, leading the race at one stage, while Triumphs finished second and third – the race was won by Dick Mann on a Honda CB750. In the following year, Mann was entered on a Rocket 3 and was victorious.

SERIES PRODUCTION

Initially, all the Triumph and BSA triples were exported; it wasn't until 1969 that they appeared on the domestic market. By then, however, Honda had announced its CB750 four, which, together with the largely unpopular styling of the British machines, seriously hampered the sales of the triples – even though the racers were cleaning up. Had the bikes been available a couple of years earlier, they would have been a full three years ahead of the CB750 and, without the competition, would have been more successful as street bikes.

1971 BSA/TRIUMPH FORMULA 750 RACER SPECIFICATIONS

Engine Air-cooled, overhead-valve, across-the-frame, three-cylinder (BSA, inclined; Triumph, vertical)

Displacement 740cc

Bore and stroke 67.0x70.0mm

Ignition Triple contact breakers, coils, battery

Gearbox Five-speed (Quaife), foot-change

Final drive Chain

Frame All-steel, full-cradle (Rob North)

Dry weight 165kg (364lb)

Maximum power 84bhp at 8,500rpm

Top speed 273km/h (170mph)

DUCATI MACH I

In the aviation world, "Mach 1" refers to the speed of sound. When testing a Ducati Mach 1 for the British journal *Motor Cycle News*, Pat Braithwaite wrote, "In a Lightning jet fighter, Mach 1 (plus a bit) takes you through the sound barrier. The new-for-1965 Ducati Mach 1 takes you through the ton barrier. And it's only a two-fifty."

ITALY V. JAPAN

Ducati's Mach 1 and Suzuki's Super Six were the fastest 250 cc street bikes of the 1960s, with a speed potential of 160 km/h (100 mph). The Ducati introduced countless would-be racers to the sport, offering a performance that was not far short of genuine, over-the-counter racers, such as the Aermacchi Ala d'Oro and Bultaco TSS.

Ducati gained the advantage over Suzuki by putting the Mach 1 into production first. In Britain, it was introduced in September 1964, at the London Earls Court motorcycle show, where it caused a sensation.

CAREFULLY HONED AND DEVELOPED

The work of Fabio Taglioni, the Ducati Mach 1 was not a brand-new design; rather, it had been carefully honed and developed from the existing 250 cc Diana (which had appeared in the spring of 1961). Both machines featured the Italian engineer's well-tried, overhead-camshaft, unit-construction, single-cylinder engine, with bevel drive to the camshaft and wet-sump lubrication.

The Mach 1 had a five-speed, close-ratio gearbox (the Diana had only four speeds), a three-ring, forged Borgo racing-type piston (with a 10.5:1 compression ratio), larger valves, a high-lift camshaft and a 29 mm Dell'Orto SS1 29D racing carburettor with separate float chamber.

Other notable features included full-width, aluminium brake hubs, a frame that used the engine as a fully stressed member, rear-set foot controls, clip-on handlebars, a one-off curved kickstart lever, a racing-style seat (only on the original 1964 series – later Mach 1s had a conventional dualseat), and an eye-catching red and silver paint job. Among the optional extras, at additional cost, were a rev-counter and aluminium wheel rims.

A RACER FOR THE ROAD

In essence, the Mach 1 was very much a road racer – it performed, handled, and braked like a pure-bred grand prix bike. As practical, every-day transport, however, the Suzuki won hands down. With the T20 Super Six, the Japanese company had taken a giant step forward, offering such technical innovations as pumped lubrication and six speeds. The machine became the forerunner of a range of civilized, high-performance, Japanese twin-cylinder lightweights.

left Rider's view of the Mach 1. Note the optional Veglia racing rev-counter.

facing page:

top Classic-machine racer John Hynes winning on his Mach 1 at Brands Hatch, summer 1999.

bottom A 1964 brochure for Ducati's sensational Mach 1, the first series-production 250 capable of 160 km/h (100 mph).

DUCATI
250
MACH/1

4 stroke
250 cc.

Timing by O.H.C.

Gearbox: 5 speeds.

Maximum speed approx.
Kms/h 150 (Ml./h 93) in
normal position - Kms/h
170 (Ml./h 106) in lower-
ed position.

Fuel consumption for 100
Kms lt. 4 (60 Ml./1 U.S.
gal. - 70 Ml./ 1 imp. gal.).

ENGINE - Single-cylinder - Bore mm. 74 (2.9134 inch) - Stroke mm. 57.8 (2.2756 inch) - Displacement cc. 248.589 (15.1606 cu. inches) - Compression ratio (2.756 inch) - Displacement cc. 248.589 (15.1606 cu. inches) - Compression ratio 10 : 1 - Timing by O.H.C. valves inclined 80° - Maximum output at the driving shaft, CV 28 (27.6180 HP) - Maximum revolutions per minute, 9,500 - Air cooling - Lubrication: forced by gear pump - Oil sump in crankcase - Ignition by distributor - 3-lights lighting recharged by flywheel alternator and static regulator of current - Battery headlamp - Tail light with Stop - Horn - Transmission: from gearbox to engine by gears; from gearbox to wheel; by chain with special runnable drive - Gearbox 5 speeds, gears in constant mesh - Pedal control in unit with the engine - 5 speeds, gears in constant mesh - Pedal control in unit with the engine - Starting by articulated pedal - Clutch: multi-plate discs running in oil-bath.

FRAME - Highly resistant steel tubing - Built on very smart lines - Front suspension: telehydraulic fork with steering dampers - Rear suspension: swinging fork with adjustable hydraulic fork with uncovered springs shock absorbers - Wheels: spoke type, chromium steel rims with normal profile 18" x 2 1/4 - Front wheel supplied with removable axle - Brakes: expanding type: front, hand operated: rear, foot operated: Drum diameter: front, 180 mm. (7.0865 inch); rear, 160 mm. (6.2992 inch) - Tyres: front, 2.50-18 ribbed; rear, 2.75-18 reinforced with block treads.

Weight (unladen) .... Kg. 116 (lb. 255.735)
Oil sump holds approx. ...... lt. 2.400 (U.S. gal. 4.409)
Fuel tank holds .... lt. 16 (imp. gal. 3.5196 - U.S. gal. 4.2267)

- On request it is also supplied with a more comfortable saddle, of a non sport model and with high handlebar.

MOTORCYCLE

DUCATI MECCANICA S.p.A. - CAS. POST. 313 - PHONE: 49.16.91 - BOLOGNA (ITALY)

PRINTED IN ITALY

DUCATI MACH 1 SPECIFICATIONS

Engine Air-cooled, overhead-camshaft, single-cylinder

Displacement 248cc

Bore and stroke 74.0x57.8mm

Ignition Coil, battery

Gearbox Five-speed, foot-change

Final drive Chain

Frame All-steel, welded

Dry weight 116kg (256lb)

Maximum power 28bhp at 8,500rpm

Top speed 160km/h (100mph)

PRODUCTION RACING

The racing of production motorcycles not only did much to sell the individual brands (especially if they won), but also helped develop the machines, particularly once factory participation began in the late 1950s. At the beginning of the 1970s, Formula 750 (later Superbike) racing evolved, challenging the established grand prix series. Yet another facet of production-bike competition was endurance racing. Subsequently, this and superbike racing achieved world championship status.

EARLY DAYS
Production racing began in the very early days of the motorcycle. For example, French Werners and Czech Laurin & Klements clashed in the 1902 Paris-Vienna event. This was a gruelling 990km (615-mile), four-day event, which involved scaling the Arlberg Pass, some 1,800m (6,000ft) above sea level, over largely unsurfaced roads.

As motorcycle development progressed, road racers fell into one of three categories: specialized grand prix (factory) racers, factory-built production bikes, and production (sports) machines. The last are considered here.

THE BOL D'OR
The beginnings of the legendary French Bol d'Or 24-hour race can be traced back to 1922 (when it lasted for 48 hours!). This first event was held at the Vaujours circuit. Since then, various venues have been used, including St Germain, Fontainebleau, Mont'lhéry, and, finally, Le Mans. It is the longest-running day-and-night event in the motorcycle racing calendar. During the 20th century, the most successful rider and motorcycle combination in Bol d'Or history was Gustave Lefevre and his Norton 500cc International. The pairing gained six victories between 1947 and 1957.

facing page A 600cc BMW R68 was used to win Australia's first 24-hour motorcycle race, at the Druitt circuit, New South Wales, in 1954.

above In the early 1970s, a production race was held during the annual international meeting at Silverstone. It featured a "Le Mans" start, where riders sprinted to their machines.

right The Norton Commando was popular for production racing. Note the high-level exhaust system.

THRUXTON

In Britain, Thruxton airfield, in Hampshire, was the scene of countless production racing battles during the 1950s–70s. The first event (lasting nine hours) took place in June 1955. By the beginning of the 1960s, the race had been set at 805km (500 miles) and had become a hotbed of competition between such marques as Triumph, BSA, BMW, and Norton. Each of the first three won races, but Norton was victorious three years running. By the beginning of the 1970s, the competitors were mounted on Norton Commandos, Honda CB750s, BSA Rocket 3s, and Triumph Tridents.

BARCELONA AND SPA

Other long-distance production races took place in Spain at Barcelona and in Belgium at Spa-Francorchamps. Kawasaki and Honda were the first Japanese manufacturers to realize

how vital these marathons could be to their development programmes. Eventually, Suzuki and Yamaha followed the same route, as did BMW, Ducati, and Laverda during the 1970s. Many of the components motorcyclists accept as standard equipment today were developed and tested in the heat of competition, including cast-alloy wheels, front and rear disc brakes, aerodynamic streamlining, multi-valve engines, twin headlamps (for maximum night vision), and aircraft-type, quick-action filler caps. Other race-inspired features are monoshock rear suspension, aluminium frames, inverted front forks, and exotic materials for producing strong and very lightweight engine components for increased performance.

There is no doubt that production racing, in all its forms, has greatly assisted motorcycle development. Moreover, because the machines are based on series-production bikes, the ordinary rider has benefited from that improvement.

DISC BRAKES

The first time a disc brake was seen on a motorcycle was in November 1955, when the famous Maserati automobile company displayed a new 246 cc, overhead-valve, single-cylinder sports model at the Milan show. The brake was mechanically operated, rather than hydraulic, and the machine went on sale a few months later. This ensured that Maserati made it into the history books, although, in the long term, the marque did not make the grade as a bike builder.

Then, in 1961, the American Midget Motors Autocycle Company offered a disc brake as an optional extra on one of its scooter-like machines.

LAMBRETTA SETS THE PACE

It was the Italian Innocenti company's Lambretta TV175 Series 3 scooter, however, that became the first powered two-wheeler to be mass-produced with a disc brake as standard equipment. Acting on the front wheel only, it was a mechanically operated device, controlled by cable from the handlebar lever. The brake comprised a full-circle friction pad that was clamped against the rotor. The latter was prone to uneven wear and warping if overheated by prolonged use.

Obviously, for higher-performance motorcycles, a fresh approach was required. This came about in two ways. The first was developed by the Italian Campagnolo company, which produced a mechanically operated, caliper-type device for the mid-1960s MV Agusta 600 four luxury roadster. It was also tested on the Benelli 250 four-cylinder grand prix racer.

A HYDRAULICALLY OPERATED DISC BRAKE

The second variant comprised a disc (usually in cast iron) with a caliper that was operated hydraulically by a handlebar-mounted master cylinder; this proved more effective and, today, is standard equipment on virtually all motorcycles. Early examples employed two circular pistons within the caliper, which were clamped on to the brake disc by hydraulic pressure. Over the years, the number of pistons has increased, and some systems have pistons of different diameter within the same caliper.

ABS

BMW was the first company to offer an anti-lock braking system (ABS) for motorcycles. Initially, this was offered as an option only on the K100 during the 1988 model year, having been unveiled at the Cologne motorcycle show in 1986.

BMW's experiments with ABS had begun in 1978, when one of its twins was fitted with an adaptation of the system developed for BMW cars. This required extensive modification to work with the bike's brakes and was very cumbersome; it was soon abandoned. After experimenting with a hydro-mechanical system, BMW chose the electronic/hydraulic anti-lock device developed by FAG Kugelfischer, as it was easily connected to motorcycle brakes and was effective in use. It was based on impulse sensor gears, which were carried on the inside of the front and rear discs. An improved and much lighter system, ABS II, arrived during the early 1990s.

facing page Peter Williams was in a class of his own during the 1973 Isle of Man TT. He won the premier Formula 1 event by more than three minutes, averaging 169.7 km/h (105.47 mph) on his Norton. The machine had disc brakes on both wheels.

right An Italian Scarab disc brake on a 1976 MV Agusta 750 America. The caliper contained twin pistons, while the solid disc was of cast iron.

below A rear-mounted disc brake with Italian Brembo caliper and semi-floating drilled disc.

below right A British Lockheed twin disc brake assembly with twin-piston calipers.

7 SUPERBI

KES

below The 981 cc, double-overhead-camshaft, three-cylinder Laverda Jota created history in 1976 by becoming the first series-production road bike to achieve 224 km/h (140 mph).

facing page Following Mike Hailwood's comeback victory on a 900 cc Ducati V-twin in the 1978 Isle of Man TT, the Italian marque introduced the MHR (Mike Hailwood Replica) for 1979. It was Ducati's top-selling model during the early 1980s.

SUPERBIKES

In 1969, Honda introduced its ground-breaking four-cylinder, four-stroke CB750, and the age of the superbike had begun.

Although revolutionary for a series-production roadster, the across-the-frame engine configuration of the CB750 had been employed by the Italians, and Honda itself, to win grand prix races and world championship titles. Even so, given the general conservative nature of motorcyclists at the time, Honda had made a bold move. But it worked, reinforcing the Japanese company's position as world leader in the industry.

At first, Honda's Japanese rivals attempted to compete with high-performance two-strokes, among them the Kawasaki H1 Mach III, Suzuki GT750, and Yamaha YR3. It was not until the arrival of Kawasaki's 903 cc, double-overhead-camshaft Z1

four in 1973, however, that the CB750 finally met its match. Producing 82 bhp, the Z1 had a maximum speed of 210 km/h (130 mph). Like the CB750, it sported a disc front brake, electric starter, five-speed gearbox and full instrumentation.

By 1977, Kawasaki had increased the bike's capacity to 1,015 cc, creating the Z1000. Like Honda, with its GL1000 GoldWing flat-four and CBX1000 six, however, this was not enough, and in 1978 the Z1300 arrived, with a liquid-cooled, across-the-frame, six-cylinder engine and shaft final drive.

The other Japanese marques were much slower to follow the four-stroke fashion. Suzuki launched its double-overhead-camshaft GS750 in 1977, while Yamaha's XS750 triple appeared a year later.

THE EUROPEAN RESPONSE

BMW's long-awaited replacement for its Earles-fork "boxer" twins was introduced at the Cologne motorcycle show in late 1969, in the shape of the Stroke 5 series, one of which was the R75 750. Then, in October 1973, the R90S 900 sportster arrived, followed by the fully faired R100RS 1000 in September 1975. All of these machines proved popular, although their high prices restricted sales.

During the 1970s, however, many motorcyclists looked to Italy for flair in styling and design, and they were not disappointed. Almost at the drop of a hat, the likes of Benelli, Ducati, Laverda, Moto Guzzi, and MV Agusta produced a glittering array of exotic superbikes.

Each of these marques was able to retain its identity with different engine configurations. Benelli chose the multi-cylinder route, with across-the-frame fours and sixes; Ducati, the 90-degree, inline V-twin; Laverda, parallel twins and across-the-frame threes; Moto Guzzi, 90-degree, transverse V-twins (with shaft final drive); and MV Agusta, a double-overhead-camshaft four based on its world championship-winning racing bikes.

Meanwhile, the British and American manufacturers continued to offer largely the types of machine that they had sold in the previous decade. Specialist producers, however, such as Dunstall, Rickman, and Bimota, found a niche market selling re-framed Japanese multis.

HONDA CB750

Honda's 1969 CB750 is one of the most important motorcycles of all time. Like Edward Turner's 1937 Triumph Speed Twin, which was produced (and copied) for more than three decades, the Japanese machine had a profound influence on future design.

FIRST NEWS

The first news of an impending Japanese onslaught on the high-performance, sporting motorcycle market – then dominated by British singles and twins – came in 1964, when a journalist reported seeing an entirely new, 500cc-class, vertical twin under test while visiting the Honda factory. Up to that time, many in the industry believed that the Japanese would leave production of larger-engined bikes to others, concentrating instead on the lightweight sector. The new machine, later identified as the CB450 (with double overhead camshafts and torsion-bar valve springs), was the very motorcycle the British had hoped the Japanese would never build. In fact, it was only the first of many.

THE CB750 ARRIVES

Although there was nothing new in an across-the-frame four (the Italians had been winning races for years with the same format, as had Honda itself), the CB750 was the world's first truly modern, mass-produced, four-cylinder motorcycle. That is why it made such an impact.

The CB750 was powered by an air-cooled, single-overhead-camshaft engine displacing 736 cc (61.0 x 63.0 mm bore and stroke), with an integral five-speed gearbox and wet, multi-plate clutch. There were two valves per cylinder, and the camshaft was driven by a centrally located chain. Other notable features included a quartet of carburettors and four exhaust pipes. The frame was a conventional tubular-steel device with a gaitered, telescopic front fork and twin-shock, swinging-arm, rear suspension.

INNOVATIVE FEATURES

More innovative was the 12-volt electrical system with push-button starting, hydraulically operated, single front disc brake (the rear was a drum), and large, matching instruments.

The flexible engine (which produced 67 bhp at 8,000 rpm) and large, comfortable dualseat made the big Honda a practical all-rounder. Not only could it achieve 200 km/h (124 mph), but it was also a good two-up tourer.

The single-cam model ran until 1978, when it was replaced by a double-overhead-camshaft, 16-valve version, the 95 bhp CB900, which could reach 217 km/h (135 mph).

The most significant achievement of the CB750, however, was the effect it had on Honda's rivals. Eventually, Kawasaki, Suzuki, and Yamaha all offered similar models and, thus, the "Universal Japanese Motorcycle" (UJM) was born.

1969 HONDA CB750
SPECIFICATIONS

Engine Air-cooled, single-overhead-camshaft, eight-valve, across-the-frame, four-cylinder

Displacement 736cc

Bore and stroke 61.0x63.0mm

Ignition Coil, battery

Gearbox Five-speed, foot-change

Final drive Chain

Frame All-steel, tubular construction

Dry weight 218kg (480lb)

Maximum power 67bhp at 8,000rpm

Top speed 200km/h (124mph)

ROTARY TECHNOLOGY

For much of the 1960s and well into the 1970s, the Wankel rotary engine appeared to be a serious competitor – and even a potential replacement – for the conventional, reciprocating piston engine. The Wankel had many apparent advantages: low weight per horsepower, tolerance of low-grade and unleaded fuels, and a smooth, vibration-free power delivery. In many ways, the basic design was a brilliant example of mechanical engineering. It was destined to suffer many setbacks during the course of its development, however, not least the blight of heavy consumption in a fuel-conscious age.

DR FELIX WANKEL

The rotary engine was the brainchild of the gifted German engineer Dr Felix Wankel, who had experimented during World War II with disc valves for torpedo engines. In the 1950s, he worked with NSU to apply his theories of rotary technology to a simple supercharger-compressor. This enabled a 49cc, NSU moped engine to propel a streamlined motorcycle to a record-breaking 196km/h (121.9mph).

Although generally referred to as a rotary engine, the Wankel has a combustion chamber shaped like a fat-waisted figure-of-eight, described as epitochoidal. In place of a normal piston, it has a rotor that spins eccentrically within the chamber. This "rotary piston" is of triangular shape and connected to the central power shaft by gearing. It is supported on eccentric bearings that allow it to rotate while keeping its three tips in contact with the sides of the chamber. Gas sealing strips are fitted at the tips of the rotor, a feature that proved to be the

engine's Achilles' heel in the early days. Subsequently, ceramic materials were developed to overcome the problem. In fact, the gas sealing difficulties suffered by the Wankel-engined RO80 car were instrumental in NSU's financial collapse at the end of the 1960s.

WANKEL-ENGINED MOTORCYCLES

Despite the fact that there have been several Wankel-engined cars – the Japanese Mazda company has sold a series of rotary sports models, for example – only four motorcycles have ever reached production: the Hercules W2000, the Suzuki RE5, the Norton Rotary, and the Van Veen OCR2000. Other companies, including Honda, Kawasaki, and MZ, have built prototypes.

The first production machine to appear was the German Hercules, which went on sale in 1974. This was equipped with a Sachs single-rotor, air-cooled engine of 294cc working chamber displacement, producing 32bhp at 6,500rpm.

Next came the Suzuki RE5; its single-rotor engine displacing 497cc and delivering 62bhp. This proved a costly financial failure, however, due in part to its ungainly appearance and rushed development.

The awesome 100bhp, Dutch-made Van Veen entered the market in 1976. Its high cost and limited appeal, though, meant that it remained a rarity.

The first production Norton Rotary was the police Interpol II of the early 1980s. Then came the Classic, F1, and the very successful racing models.

ITALIAN EXOTICA

Although Honda had stunned the motorcycle world with the launch of its CB750 four – the world's first superbike – during the 1970s, Italian machines were renowned for their styling and performance. In response to the Japanese invader, the likes of Ducati, Laverda, Moto Guzzi, and MV Agusta produced a glittering array of exotic superbikes.

HOW IT ALL BEGAN

Although Italy had long been a producer of mouth-watering, high-performance, sporting lightweights, Moto Guzzi and Gilera alone had built larger motorcycles, and they had been only 500cc singles, such as the Guzzi Falcone and Gilera Saturno. Italy's first really-large-capacity bike arrived during the mid-1960s, in the shape of the Moto Guzzi V7. Few motor-

cycles can have had a stranger evolution: its origins – or at least those of its 700cc, overhead-valve, 90-degree, transverse V-twin engine – lay in a contract with the Italian military to supply a go-anywhere tracked vehicle.

MOTO GUZZI

Work on the prototype of what would become the V7 motorcycle began in early 1964, the first civilian model appearing at the Milan show in late 1965. It was well-received, and orders soon began to flood in. In 1971, Lino Tonti created the sensational V7 Sport, with a 750cc engine and very sporting styling. A host of variants followed throughout the decade, with 750, 850 and 1,000cc engines. The most revered models were the V7 Sport (1971–74) and 850 Le Mans 1 (1976–77).

facing page:

top The 1971 four-cylinder MV Agusta 750S was an impressive piece of machinery; a high price ensured exclusivity.

bottom During the 1970s, Benelli built such multi-cylinder models as the 750 Sei, 504 Sport and 354 Sport (shown), Their engines aped Honda's.

right Moto Guzzi produced the sensational V7 Sport. It featured a 750cc, V-twin engine, five-speed gearbox and massive, double-sided, front drum brake.

below Laverda's 750SFC (1971–76) was a hand-built sports/racing machine. This 1975 model was equipped with electronic ignition.

DUCATI

In the summer of 1970, Ducati entered the superbike stakes with the 750GT, which had been designed by Fabio Taglioni and entered production in early 1971. This featured a 90-degree, overhead-cam, V-twin engine with integral five-speed gearbox. Its layout resulted in a long wheelbase but, despite this, handling and roadholding were of the highest order. The 1973 750SS Desmo was the first production bike to be equipped with triple disc brakes (two at the front, one at the rear).

An 864cc version (known as the 860) made a winning debut in the 1973 Barcelona 24-hour race, and a production version, the 900SS, arrived for 1975. Mike Hailwood's famous 1978 TT victory on a Ducati led to the 1979 MHR (Mike Hailwood Replica).

LAVERDA

Laverda had built 650cc (1966) and 750cc (1969) vertical twins, and in 1971, the limited-edition 750 SFC sports/racer was put on sale. It was the 3C and later Jota, however, with their three-cylinder, 1,000cc engines, that really created the Laverda legend – the Jota was the first (1976) production motorcycle to reach 225km/h (140mph).

MV AGUSTA

MV Agusta built small batches of its exclusive and extremely expensive, double-overhead-camshaft, across-the-frame, four-cylinder models during the 1970s, notably the 750S, 750 America, and, finally, 850 Monza. They were rare sights then and subsequently became much sought after.

BMW R100RS

BMW's long-awaited replacement for its Earles-fork "boxer" twins series (introduced in 1955) made its first appearance at the Cologne motorcycle show in September 1969. There were three versions: the R50/5, R60/5, and the R75/5. The engine remained an overhead-valve, horizontally opposed twin (enlarged to 750cc) with shaft final drive, but there were many new features, such as an electric starter, flashing indicators, a five-speed gearbox, and a telescopic front fork.

THE R90S

In October 1973, BMW made its next move, with the R90S, dubbed "Germany's sexiest superbike" by the press. The choice of the Paris show for its debut was an obvious move, since at that event 50 years before, almost to the day, BMW had launched its very first bike, the R32.

In its styling – by Hans Muth – the R90S represented a milestone in the company's history. It featured a dual "racing" seat, a fairing cowl, twin hydraulic front disc brakes, and an exquisite, airbrushed, custom paint job for the bodywork, which ensured that no two machines looked the same.

ENTER THE R100RS

The Cologne show in September 1976 heralded the Stroke 7 series, which included a new concept for BMW (and the industry as a whole) – the fully faired R100RS.

A major feature of the R100RS was its comprehensive fairing. To achieve the optimum shape, BMW hired (at considerable cost) Pininfarina's wind-tunnel in Italy. The resulting streamlined bodywork not only added additional speed, but also provided the rider (and passenger) with a level of weather protection previously undreamed of by earlier generations. The 980cc (94.0x70.6mm bore and stroke) engine was the most powerful development of the two-valves-per-cylinder, pushrod flat-twin in a road-going machine, giving a genuine 70bhp. This and the comprehensive fairing gave a maximum speed approaching 208km/h (130mph).

THE FINEST SPORTS/TOURER EVER?

Many enthusiasts consider the original 1977 R100RS to be the finest sports/tourer ever built. There have been faster machines, even bikes that handled better, but none to match the all-round abilities of the R100RS. Another important feature was the 24-litre (5.25-imperial-gallon; 6.5-US-gallon) fuel tank, which provided a range of almost 480km (300 miles) if the bike was ridden gently.

The other models in the Stroke 7 series were the R100/7, the basic unfaired version; the R100S, successor to the R90S; the R100RT, with its "barn-door" fairing; and the R75/7 (later replaced by the R80/7).

The R100RS was dropped following the launch of the K Series, but was subsequently re-introduced in response to public demand. However, many people considered the re-issued bike an inferior machine because its power output was reduced to 60bhp.

left Introduced in 1976, the BMW R100RS featured a comprehensive fairing that had been developed in the wind-tunnel belonging to famous Italian coachbuilder Pininfarina.

facing page To many, the R100RS represents the ideal sports/touring motorcycle, with its excellent rider protection, powerful twin-cylinder engine, shaft final drive, and secure handling.

1977 BMW R100RS SPECIFICATIONS

Engine Air-cooled, overhead-valve, transverse, flat-twin

Displacement 980cc

Bore and stroke 94.0x70.6mm

Ignition Coil, battery

Gearbox Five-speed, foot-change

Final drive Shaft

Frame All-steel, full duplex

Dry weight 210kg (462lb)

Maximum power 70bhp at 7,250rpm

Top speed 204km/h (127mph)

HARLEY-DAVIDSON

By the mid-1960s, Harley-Davidson was all that remained of a once great American motorcycle industry. Although it had entered into a transatlantic partnership with the Italian Aermacchi company in 1960, as far as large-displacement motorcycles were concerned, Harley-Davidson had stuck rigidly to its long-running, V-twin engine format.

In 1949, Harley-Davidson had introduced telescopic front forks. Then, in 1952, it switched to a foot gear-shift and a hand-operated clutch. The next big step came in 1958, with twin-shock rear suspension, celebrated by the company's top-of-the-range Duo-Glide.

THE ELECTRA GLIDE

In 1965, one of Harley-Davidson's most famous and revered models was introduced – the Electra Glide. This was the first of the company's bikes to be equipped with an electric, push-button starter, hence the "Electra" tag. The Harley marketing men called the Electra Glide "King of the Highway". And that's exactly how the rider felt, sitting astride all that glitter and excess. The "Glide" stood out in a crowd in the mid-1960s; later models continued to do that, even 40 years later.

A COMPETITOR

The Electra Glide cornered the market for such bold, luxurious, and expensive pieces of metal, at least until Honda introduced

its GoldWing a decade later. Although the Harley and Honda appeared to be similar bikes, under the skin they were as different as chalk and cheese.

The GoldWing was a super-sophisticated, high-tech "Flash Gordon" to the Electra Glide's brash and up-front "Dolly Parton". Although far superior technically, the Japanese bike didn't have that magic name on the tank, which is what counted to many buyers.

As one journalist put it: "I don't know about you, but give me a piece of history to a piece of plastic any day!"

TECHNICAL DETAILS

When it first entered production, the Electra Glide had a 1,207 cc (87.0 x 100.0 mm bore and stroke) version of the familiar air-cooled, overhead-valve, V-twin engine, with a four-speed, foot-change gearbox. It remained that way until the arrival of the EVO (Evolution) engine in 1983. In Harley speak, it was as big an advance over the preceding Shovelhead as the Shovelhead had been over the Panhead.

Essentially, the new V-twin was an alloy-barrelled version of the Shovelhead engine, with a lighter valve-train and better breathing. It was also considerably lighter than the previous "iron" engine and produced more power. At the same time, the capacity was increased to 1,340 cc (87.3 x 100.8 mm bore and stroke), but it was still air cooled.

The 1968 Electra Glide is unique in all cycling!

Introduce yourself to a one-of-a-kind experience. Ride the new Electra Glide. In this one motorcycle, you'll get the precise engineering of a formula racer and the handcrafted luxury of a custom roadster. Here too, are the stamina, balance and ride that have made the 1200 cc Electra Glide the world's foremost cycle. Now add electric starting, new instrumentation and futuristic styling. This is Electra Glide for 1968. On display right now at the Harley-Davidson dealer nearest you. A limited edition of unlimited excellence. Stop in for a test ride soon.

HARLEY-DAVIDSON

MOTOR HARLEY-DAVIDSON CYCLES

facing page By the mid-1960s, Harley-Davidson was all that remained of the once-great American motorcycle industry. Although its massive V-twin machines were old-fashioned, they retained a loyal following.

above The Electra Glide was launched in 1965 and dubbed "King of the Highway".

left Another of Harley's favourite descriptions was, "The great freedom machine". This picture says it all.

HONDA GOLDWING

Making its world debut at the Cologne motorcycle show in September 1974, Honda's GL1000 GoldWing reflected a radical new take on the touring motorcycle.

THE COMPETITION

Although sold in Europe, the GoldWing had been created for the US market. When designing the bike, Honda engineers had taken the Harley-Davidson as the competition. Despite its elderly V-twin engine, the Harley was big (in 1,200cc form), strong and powerful enough to carry a rider and passenger – plus luggage – over the vast distances found in the USA.

The other serious contender for the market was the BMW. The history of the German flat-twin could be traced back through decades, but, compared with the American bike, it had the advantage of shaft final drive. In the USA, however, the BMW was not considered a true tourer at the time, since the largest engine size offered was 750cc (later increased to 900 and, finally, 1,000cc).

A FLAT-FOUR

After studying a number of different layouts and engine sizes, the Honda engineers settled on a liquid-cooled flat-four, with a single overhead camshaft for each cylinder head, driven by toothed Gilmer belts rather than chains. There were four constant-vacuum (CV) carburettors, a five-speed gearbox, and shaft final drive.

Also, there was a mechanical fuel pump, because the tank was mounted beneath the dualseat. The tank-shaped bodywork over the engine was actually a dummy full of electrical equipment – and air!

A ONE-LITRE CAPACITY

The 1,000cc (72.0x61.4mm bore and stroke) engine had a 9.2:1 compression ratio and produced 80bhp at 7,000rpm. Its cylinders were staggered, being farther forward on the off-side (right) than the nearside. The crankshaft was laid out so that the front two pistons moved in opposition to the rear pair for balanced operation.

As in most other Honda multi-cylinder designs, the crankcases were split horizontally. Power was transmitted from the crankshaft to the gearbox by an inverted-tooth chain. There was also a conventional duplex chain, from the rear of the clutch drum to the two oil pumps and single water pump positioned low in the crankcase. The lubrication system was of the wet-sump type.

The GL1000 weighed 259kg (571lb), but the low centre of gravity provided good handling.

LARGER ENGINES

By the beginning of the 1980s, the GL1000's displacement had risen to 1,180cc. In 1987, the GL1500 arrived – with a six-cylinder engine – and, at the beginning of the 21st century, capacity was increased again, to 1,800cc. The original GoldWing had been "naked", but from the late 1970s, most examples came fully equipped with a large fairing and a comprehensive range of equipment.

1975 GL1000 SPECIFICATIONS

Engine Liquid-cooled, single-overhead-camshaft, flat-four

Displacement 1,000cc

Bore and stroke 72.0x61.4mm

Ignition Battery/coil

Gearbox Five-speed, foot-change

Final drive Shaft

Frame All-steel, full-duplex

Dry weight 259kg (571lb)

Maximum power 80bhp at 7,000rpm

Top speed 201km/h (125mph)

above In 1987, the first six-cylinder GoldWing appeared, being known as the GL1500. The displacement was increased to 1,800cc at the beginning of the 21st century.

left At the beginning of the 1980s, the engine size of the GoldWing was increased to 1,180cc, but it remained a four.

facing page Introduced toward the end of 1974, Honda's GL1000 GoldWing had been designed largely for the US market. It had a 1,000cc, liquid-cooled, single-overhead-camshaft, flat-four engine.

GoldWing

SIX CYLINDERS

Founded in 1911, the Italian Benelli marque has been a leader in the design and construction of multi-cylinder engines. Long before the Japanese made four-cylinder bikes commonplace, Benelli had created a supercharged, liquid-cooled, double-over-head-camshaft, 250 four grand prix racer. This exciting machine was unveiled in 1939 but, in mid-1940, Italy entered World War II and the machine was "mothballed". Subsequently, Benelli suffered another blow when the FIM banned superchargers postwar.

Two decades passed before Benelli re-entered the fray with another four, in mid-1960. Also a grand prix bike, it was a totally new, air-cooled design. In 1969, Kel Carruthers became 250cc World Champion on the Benelli four (the last four-stroke to win the title).

750 SEI

In the early 1970s, the Argentinean industrialist Alejandro de Tomaso acquired the Benelli factory and set about rebuilding its reputation. To this end, he stunned the motorcycle world in 1972 with the launch of the first road-going, six-cylinder machine, the 750 Sei.

A blatant attempt by de Tomaso to "out-super" all super-bikes, the new Benelli was a typical Italian engineering creation. From the land of Maserati, Lamborghini, and Ferrari, the single-overhead-camshaft straight-six seemed a logical product. Like those exotic cars, the 750 Sei was expensive but, unlike them, it didn't benefit from amazing styling. In fact, with the exception of the six megaphones crowding the rear wheel, its appearance owed more to Tokyo than Milan. It was

two years before limited production began in mid-1974. The 747cc (56.0x50.6mm bore and stroke) engine produced 71bhp at 8,500rpm, giving a 190km/h (118mph) top speed.

The 750 was succeeded in 1979 by the 900 Sei, the increase in engine size being achieved with bore and stroke dimensions of 60.0x53.4mm. The new bike addressed the gearbox and crankshaft weaknesses of the original and, with 80bhp and 209km/h (130mph), it was a superior machine.

CBX

All that counted for little, however, because long before the 900 Sei arrived, Honda had muscled in with its own six, the mighty CBX.

Like Benelli, Honda had grand prix experience, but unlike the Italian marque, this included six-cylinder bikes, in the shapes of the 1960s 250 and 297cc world championship-winning machines ridden by Mike Hailwood and Jim Redman. The advanced design of the CBX drew on this vast experience. For example, it benefited from seven main bearings and four valves per cylinder (the Benelli had two), while the 1,047cc (64.5x53.4mm bore and stroke) engine also featured double overhead camshafts and a compression ratio of 9.3:1. It produced 105bhp at 9,000rpm, good enough for a little more than 224km/h (139mph) when tested by *Motorcycle Weekly*.

Other features of the original CBX Super Sport included twin-shock rear suspension. In 1981, the CBX1000C, with monoshock rear suspension and fairing, was offered. Total CBX production amounted to 30,000 so, although Benelli was first in the market, Honda won commercially.

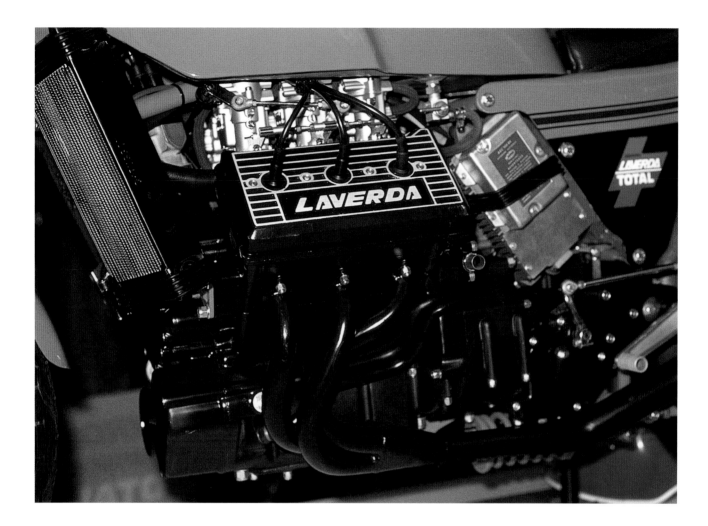

above In the late 1970s, Laverda displayed a six-cylinder prototype. It had a 995.89cc (65.0x50.0mm bore and stroke), liquid-cooled, double-overhead-camshaft, 90-degree V6 with four valves per cylinder.

right Unfortunately, the Laverda never made it into production.

facing page:

left Honda's mighty CBX arrived in 1977, featuring 1,047cc, double overhead camshafts, 24 valves, 108bhp, and 161km/h (100mph) in seven seconds.

right Shown as a prototype in 1972 in an attempt to outfox Japanese designers, the Benelli 750 Sei was a typical Italian luxury product.

TURBOCHARGING

Like the Wankel rotary engine, turbocharging was thought to offer an important technical advantage. Although all four Japanese manufacturers – Honda, Suzuki, Yamaha, and Kawasaki – put turbocharged bikes into production, ultimately, the concept did not prove worthwhile.

HOW IT WORKS

The purpose of a turbocharger is to make use of the large amount of energy carried away by exhaust gases. This energy, readily available when an engine is under load and exhaust flow is high, is employed to force a greater volume of air through the engine, effectively increasing its displacement. Thus, turbocharging delivers extra power from an engine in the mid- and high-rpm ranges.

While other manufacturers built prototypes and small batches of turbocharged machines, only the Japanese giants attempted to exploit the turbo for series-production bikes. They applied the technology in different ways, however, from the simple to the over-complex, with mixed results.

THE DESIGNS

Suzuki and Yamaha mounted turbos behind their four-cylinder engines, which made routing the exhaust pipes relatively easy, but increased response time because the turbo was a considerable distance from the exhaust ports. Honda fitted the turbo near the exhaust ports of its CX V-twin, but high in the frame, making an already high centre of gravity even worse.

COMBATING TURBO LAG

To combat turbo lag, some manufacturers (including car makers) used a reed intake valve to bypass the turbo until boost pressure surpassed atmospheric pressure. The intention was to let the engine run efficiently without boost, but this was only partly successful.

Of the Japanese marques, Kawasaki was the most successful in reducing turbo lag. To eliminate the low-rpm bypass employed on early prototypes, its engineers shortened the induction path as much as possible by placing the air filter near the engine sprocket. This was beneficial in four ways: first, it minimized the time between the exhaust gas leaving the ports and beginning to drive the turbine; second, it reduced the loss of heat energy in the turbo system; third, it produced a relatively low centre of gravity; and, fourth, it insulated the rider from heat in the system.

A WASTE-GATE

The Kawasaki turbocharging system was protected by a waste-gate, which allowed the exhaust gases to pass around the turbo if boost pressure reached 560mm (22in) Hg. As a further safety measure, the digital fuel injection (DFI) system shut off the fuel supply to the engine if boost reached a dangerous level.

Kawasaki also pioneered the use of liquid crystal display (LCD) on its 1984 750 Turbo model to indicate boost pressure – a first in motorcycling.

XJ650T

- Yamaha-Turbo-System mit YICS.
- Aerodynamische Vollverkleidung.
- Computergesteuertes Monitor-System.
- Kardan-Antrieb.
- Luftunterstützte Telegabel.
- Mehrfach verstellbare Federbeine hinten.

- Système de Turbo Yamaha complété par l'YICS.
- Carénage intégral aérodynamique.
- Système de contrôle par ordinateur.
- Transmission par cardan.
- Fourche hydropneumatique.
- Suspension arrière entièrement réglable.

- Yamaha Turbo System with YICS.
- Aerodynamic full fairing.
- Computer Monitor System.
- Shaft drive system.
- Air-assisted front fork.
- Fully adjustable rear suspension.

- Sistema Turbo Yamaha con YICS.
- Carena aerodinamica.
- Sistema di controllo computerizzato.
- Trasmissione finale a cardano.
- Forcella anteriore pneumatica.
- Sospensione posteriore regolabile.

- Beim Yamaha-Turbosystem findet der kleinste Turbolader der Welt Verwendung. CDI-Zündung und Membrane sorgen für direktes Ansprechen des Motors.
- Le système Turbo Yamaha utilise le turbo-compresseur le plus petit du monde. L'allumage ainsi que l'admission de gaz frais supplémentaires contrôlés par ordinateur apportent une puissance exceptionnelle dans les plus hauts régimes.
- The Yamaha Turbo System utilises the world's smallest turbo unit, computer-controlled ignition and reed valve, producing exceptional power with instant response.
- Il sistema Turbo Yamaha utilizza un gruppo turbo di dimensioni più ridotte di tutti gli altri esistenti, ha l'accensione controllata mediante computer, le valvole a lamelle ed eroga una potenza eccezionale.

facing page Exhibited at the Cologne motorcycle show in 1980, the BMW Futuro flat-twin prototype had a turbocharged, 800cc engine and horizontal, monoshock rear suspension.

above At the time of its launch in 1982, Yamaha's XJ650T was claimed to have the world's smallest turbo unit.

below The Honda CX500's turbo was mounted near the exhaust ports, but this caused centre-of-gravity problems.

SPECIALIST MANUFACTURERS

Since the advent of the modern superbike, which occurred in 1969 when Honda launched its CB750 four, the world's motorcycle manufacturers have built a seemingly never-ending stream of desirable and often innovative models.

THE SUCCESS OF THE SPECIALISTS

The major manufacturers have massive development budgets and employ the best designers, so it would appear almost impossible for the limited-production specialist builder to come up with a superior product and, just as importantly, make a profit. Against all the odds, however, this is just what has occurred in a number of cases.

The success of specialist builders is due to a variety of reasons. For a start, there have always been motorcyclists who have demanded bikes that are truly unique. Then there is the fact that the early Japanese machines were not blessed with particularly good handling, while the Italians have

produced some excellent bikes, but with appalling levels of finish. Finally, the major producers have always had to consider the increasingly vociferous "green" lobby, with the result that many series-production bikes have been almost overloaded with environmental features.

THE IMPORTANT NAMES

Among the plethora of specialist builders that have flourished in this environment are Dunstall, Seeley, Rickman, Egli, Bimota, Dresda, Münch, Moto Martin, Harris, Nico Bakker, Moko, and Magni. Of these, Bimota is probably the most famous and widely respected, in much the same way as such marques as Ferrari, Lamborghini, and Maserati in the four-wheel world.

Bimota (the name comes from the first two letters of its founders' names: Bianchi, Morri, and Tamburini) never set out to be a manufacturer, but simply to improve the handling of

facing page Bimota's 1997 500 V Due had a low-pollution, high-output, two-stroke engine, but proved a commercial failure.

left In 1985, Bimota introduced the Ducati-based DB1 in racing and road versions.

below The Magni Australia Battle of the Twins racer was built by former MV Agusta team manager Arturo Magni in the early 1990s.

bottom The 1988 Bimota YB4 superbike racer had a 20-valve Yamaha FZ750 engine.

such machines as the MV Agusta 600 and Honda CB750 by supplying replacement frame kits. The company was formed in January 1973 with the launch of its prototype racing machine, the Honda-engined HB1. Since then, Bimota has gained a reputation for innovation.

VARIABLE STEERING GEOMETRY

The innovative features that have come from Bimota have largely been the work of Massimo Tamburini and have included variable steering geometry, a space-frame (made from a large number of small-diameter tubes), and monoshock rear suspension. Also notable is the 1977 SB2 (Suzuki Bimota 2). On this design, the 18-litre (four-gallon) fuel tank was mounted below the engine, with an electric pump to lift fuel to carburettor height. Putting the tank in this position lowered the centre of gravity by some 14 percent, compared with Suzuki's own GS750 of the same era.

8 THE MOD
MOTO

ERN

RCYCLE

THE MODERN MOTORCYCLE

During the 1980s and early 1990s, the world's motorcycle industry made great technical advances. For once, Honda didn't lead; instead, the front-runners were Japan's Kawasaki and Suzuki, Italy's Ducati, and Germany's BMW.

Although Honda continued to sell the most machines in the early 1980s, it seemed to lack a coherent strategy, at least in the big-bike field. This was typified by the company's series of V4 designs, which took much longer to develop than it anticipated. The other Japanese marques were left to push the boundaries of technology.

Kawasaki was the first to put a new generation of big-bore sports bikes onto the market, in the shape of the all-new GPZ900R (Ninja in North America). Shown to the press in December 1983, it was the world's first liquid-cooled, 16-valve, across-the-frame, four-cylinder production bike. The GPZ900R was also the first to employ a lightweight, diamond frame, aluminium sub-frame, and 406mm (16in) front wheel. Another feature was Kawasaki's innovative Uni-Trak monoshock rear suspension.

Suzuki really set things alight with its "racer-for-the-road" GSX-R750 of 1985. As the *American Motor Cyclist* magazine commented in its March 1996 issue: "Contemporary sports bike history is clearly divisible into two chunks: the Dark Ages, i.e. before the 1985 GSX-R750, and the Renaissance thereafter!" It wasn't simply that the GSX-R750 was the fastest; it was a complete sports bike, with superb performance, handling, and braking wrapped in a mouth-watering package.

facing page Kawasaki's ZX-10 arrived in late 1987, replacing the GPZ1000RX. With more power and torque, and a higher speed (272 km/h; 169 mph), it assumed the mantle of world's fastest production roadster.

below The Honda CBR1000F was built between 1987 and 1999. It began life as a sports bike, but was really a comfortable sports/tourer.

Of the European manufacturers, only BMW built anything that was really new in the first half of the 1980s, when it launched its K Series range. The first of these was the four-cylinder K100, soon nicknamed the "flying brick" because of the engine shape. From this, the K100RS sports/tourer and three-cylinder 750 K75 were developed. Later still came the 16-valve K1. All the K Series machines had fuel injection, which was an industry first. BMW was also the first to offer ABS (anti-lock braking) on a production bike, albeit originally as a cost option.

Ducati was taken over by Cagiva in 1985, and the marque introduced a number of innovative designs over the following decade, many penned by Massimo Tamburini. Engine development was carried out by Massimo Bordi, who

had replaced Fabio Taglioni. The first to appear was the 750 Paso, in late 1985, but the really big news came two years later with the 851. This design was the first Ducati to incorporate four valves per cylinder, double overhead camshafts and fuel injection in the marque's 90-degree V-twin. Later, the 851 became the 888, with which Ducati won the first of many World Superbike titles in 1990.

In the superbike stakes, the first half of the 1990s was dominated by the Honda CBR 900RR FireBlade and Ducati 916. Other important developments that occurred during this time included the new 600cc super sport class; improvements in handling, comfort, tyres, and braking performance; and the prevalence of digitally controlled ignition and fuel injection systems.

KAWASAKI GPZ900R

Launched in December 1983, the Kawasaki GPZ900R was a trend setter. Sold as the Ninja in North America, it had a liquid-cooled, 16-valve, double-overhead-camshaft, across-the-frame, four-cylinder engine. Other unique features were a lightweight, diamond main frame with aluminium rear section and 406mm (16in) front wheel. Kawasaki claimed that it was the first bike to have a front fork that delivered truly progressive wheel travel. At the rear was the company's rising-rate Uni-Trak suspension.

THE ENGINE

As with the earlier air-cooled Z1, the centrepiece of the new machine was its engine. To create a second-generation superbike, Kawasaki had carried out extensive tests on several configurations, including a V4, a V6, and an across-the-frame six, but none was found to offer any significant advantage over an across-the-frame four. Consequently, the proven layout was

retained, although virtually every other aspect of the power unit was re-evaluated. The result was an all-new engine, which allowed Kawasaki – and ultimately the rest of the Japanese motorcycle industry – to take a giant step forward.

An important aspect of the 900R's engine was its compact size – it was narrower and less tall than that of the Z1. A major factor determining its smaller dimensions was a basic change in its layout. Liquid cooling meant that the cam chain could be positioned outside the cylinder bank, instead of in the centre. As a result, the wet-cylinder cooling system was both compact and highly efficient. Revving to 10,500rpm, the one-piece crankshaft was supported by five plain bearings. In addition to driving the clutch, the crankshaft also drove a compact counter-balancer, which virtually eliminated secondary vibration. The clutch was operated hydraulically, while the gearbox contained six speeds – a first on a large-capacity Kawasaki motorcycle.

facing page The Kawasaki GPZ900R was a trend setter, being the first superbike with a liquid-cooled, 16-valve, across-the-frame, four-cylinder engine.

above When launched in 1983, the GPZ900R was considered a sportster. Later, however, it became very much a sports/tourer and was often equipped with hard luggage.

right The high-tech, 114 bhp GPZ900R power unit was also used in a number of specials, including this show-winning, Harris-framed machine.

THE CHASSIS

A new, high-tensile steel frame reduced weight by about 5 kg (11 lb), compared with previous Kawasaki fours; it also lowered the centre of gravity. In the early stages of testing, standard downtubes were added to determine what stress, if any, they would counter. Results showed that the downtubes carried virtually no load, however, and they were discarded. A box-section, aluminium, rear swinging arm employed eccentric snail-cam chain adjusters.

Other features of the 900R included triple disc brakes, an anti-dive front fork, six-spoke alloy wheels, V-rated tyres, and a dry weight of 228 kg (503 lb).

Motor Cycle News recorded 254 km/h (158 mph) with the GPZ900R, while a trio of the bikes took the first three places at the 1984 Isle of Man TT. Technology moves on, however, and by 1985, the Kawasaki had been outstripped by even faster bikes, but it had been the first of the new breed.

KAWASAKI GPZ900R SPECIFICATIONS

Engine Liquid-cooled, double-overhead-camshaft, 16-valve, across-the-frame, four-cylinder

Displacement 908 cc

Bore and stroke 72.5x55.0 mm

Ignition Electronic

Gearbox Six-speed, foot-change

Final drive Chain

Frame Steel backbone, aluminium sub-frame and swinging arm

Dry weight 228 kg (503 lb)

Maximum power 114 bhp at 10,500 rpm

Top speed 254 km/h (158 mph)

ENDURO STYLE

The famous Paris-Dakar Rally, first staged for motorcycles in 1979, and the International Six Days Enduro (ISDE) – the latter replacing the long-running International Six Days Trial (ISDT) – were largely instrumental in the rise in popularity of enduro-styled bikes. These have been based on the successful works machines created by the likes of BMW, Cagiva, Gilera, Yamaha, and Honda.

PARIS-DAKAR

The first Paris-Dakar event, billed as "The World's Toughest Race", was won by Cyril Neveu on a modified Yamaha XT500 trail bike. However, BMW was the first manufacturer to really exploit the incredible 9,975 km (6,200-mile) marathon, from the French capital to Dakar in Senegal, West Africa.

Competitors in the Paris-Dakar are faced with two weeks of biking torture, as first they meet the challenges of the Atlas Mountains and then the Sahara desert, one of the most inhospitable and dangerous places on Earth. Daytime temperatures soar well above 38°C (100°F), but it can freeze at night; sandstorms can blow up out of nowhere, scouring the face and obliterating the trail. The Paris-Dakar is truly the ultimate test of both man and machine.

THE BMW EFFORT

Special versions of the BMW flat-twin were victorious in 1981 and in 1983–85. After this record of four wins, BMW suffered a poor result in the 1986 event and subsequently announced its retirement from the rally (although it did return in the late 1990s).

Herbert Auriol is the only man to have won Paris-Dakar on two and four wheels. A former French trials champion, he won in 1981 on a BMW GS800, and in 1983 on a BMW 980. (In 1992, he was victorious in the car section, with a Mitsubishi.) Former World Motocross Champion Gaston Rahier, from Belgium, won the 1984 and 1985 events on 1,000cc BMWs.

All the German machines featured long-travel suspension, two-into-one exhaust systems and single, side-mounted rear shocks. BMW, together with Yamaha, pioneered the use of massive fuel tanks to allow over 580 km (360 miles) to be covered without refuelling.

CAGIVA, YAMAHA, AND GILERA

After BMW quit, the main competitors were Cagiva, Yamaha, and Gilera. Cagiva used specially prepared, 900cc, Ducati-engined V-twins; the others used high-tech singles. The Yamaha 660 had five valves, six speeds, a long-travel front fork, and monoshock rear suspension. Gilera's RC600, which dominated its class in the early 1990s, displaced 558cc (98.0 x 74.0mm bore and stroke) and had a four-valve head, twin exhaust ports, a five-speed gearbox, and two TK carburettors.

HONDA EXP-2

Among the most technically interesting bikes ever used in the Paris-Dakar, however, are a pair of experimental 400cc Honda EXP-2 singles, which were entered in 1995. These employed active radial combustion (ARC) – a chemical self-ignition system that was related to the combustion process of the World War II German V2 rocket!

left BMW built enduro versions of its flat-twin for the Paris-Dakar Rally. These were victorious in 1981, 1983, 1984, and 1985.

facing page:
top A 1977 Ducati 125 Six Days enduro bike being put through its paces. The rear shocks could be mounted at an angle (as shown) or vertically.

bottom left The 2005 Husqvarna 610 featured monoshock rear suspension, an inverted front fork, and aluminium rear swinging arm.

bottom right During the late 1980s and early 1990s, Gilera won the 600cc class of the Paris-Dakar more than once with the liquid-cooled RC600R single.

BMW K SERIES

During the late 1970s, BMW's motorcycle sales were slipping, particularly in the USA. With warehouses full of unsold bikes, the chairman, chief engineer, and sales director all left the German company. Fortunately, their replacements, who joined in 1979, had considerable motorcycle experience.

THE SEARCH FOR A NEW IMAGE
BMW motorcycles needed a new image that encompassed more than the long-running flat-twin. In fact, one of the company's employees, Josef Fritzenwenger, was already looking for a replacement for the venerable model – something he had been doing since the mid-1970s, with little support.

Fritzenwenger's first idea was a water-cooled flat-four, but Honda's introduction of the same layout for its new GoldWing killed this off. BMW didn't want to appear to have copied one of its competitors.

COMPACT DRIVE SYSTEM
Eventually, Fritzenwenger came up with the concept of the compact drive system (CDS), after carrying out tests on a

liquid-cooled, four-cylinder engine from a Peugeot 104 car. Originally, the aluminium engine had been inclined and mounted transversely in the 104. It occurred to the BMW engineer, however, that if he laid it flat, with its crankshaft in line with the frame, it could be just what he needed.

The configuration was ideal for shaft final drive, had a low centre of gravity, and offered good accessibility. Moreover, no other bike maker built a longitudinal and horizontal inline four – or had ever done so.

THE PROTOTYPES
Various prototypes – some with three cylinders, others with four – were built, ranging from 800 to 1,300cc. Ultimately, however, two engine sizes were selected for production: the 740cc K75 and 987cc K100. Both had 67.0x70.0mm bore and stroke dimensions.

Work began on the range in May 1979. The development team, now led by Stefan Pachernagg, concentrated initially on the four, which, as the unfaired K100, appeared in October 1983. This was followed by the far more popular K100RS

facing page Dubbed the "Flying Brick" by many, the K100 employed BMW's compact drive system.

left In addition to bold styling, the K1 Super Sports model, which appeared in late 1988, was notable for having four valves per cylinder (a first for BMW's motorcycle division).

above Members of BMW's development team with their handiwork in September 1983, just prior to the launch of the K100.

sports/tourer and, finally, the LT version, which was equipped with a massive fairing.

A major innovation on the K Series BMWs was the use of Bosch LE-Jetronic fuel injection. Digital ignition was another significant advance.

THREE CYLINDERS

The three-cylinder K75 was launched in September 1985, the "S" version following in June 1986. Technically, the only major difference between the three- and four-cylinder models was the K75's contra-rotating balance shaft, which virtually eliminated vibration (a problem on the larger engine). It also had a modified R80 clutch and straight-cut gears.

The 16-valve K1 (with the K100's 987cc displacement) arrived in September 1988, further developments leading to 1,100 and 1,200cc models. In 2005, the brand-new 1,157cc, 163bhp K1200R (Roadster) and K1200S (Sport), with across-the-frame engines, were introduced.

1984 BMW K100 SPECIFICATIONS
Engine Liquid-cooled, double-overhead-camshaft, inline, flat-four
Displacement 987cc
Bore and stroke 67.0 x 70.0mm
Ignition Digital, integrated with electronic fuel injection
Gearbox Five-speed, foot-change
Final drive Shaft
Frame ubular space-type, engine used as stressed member
Dry weight 230kg (507lb)
Maximum power 90bhp at 8,000rpm
Top speed 215km/h (134mph)

SUZUKI GSX-R750

From the beginning, the GSX-R750 had been intended as a racer first, a street bike second. Its designer, Tansunobu Fujji, ensured that it met that criterion.

OIL COOLING

Initially, a working prototype of the machine was campaigned in endurance racing events during the early 1980s. One feature of this that was adopted for the production model, and which set the GSX-R apart, was oil cooling. Although this idea was not new, having been employed to help control the temperature of aero engines since the 1920s, it is important to understand the reasoning behind its use in a motorcycle.

Suzuki had opted for oil cooling to cure overheating problems encountered during development of its XN85 Turbo. Oil jets directed at the undersides of the pistons prevented the turbocharged engine from dropping molten metal into the crankcase. The same idea was used on the 1983 GSX-750EF sports/tourer to prevent the two central pistons from running hotter than the outer pair.

THE "R" ENGINE

While it would be accurate to describe the GSX motor as being air cooled with oil assistance, the emphasis was reversed on the "R" unit. Oil cooling came into its own in the new engine, which employed a second oil pump purely to remove engine heat. The primary pump, mounted on the same shaft and sharing a common housing, handled lubrication. A large oil cooler was mounted on the front frame tubes.

In answer to the question why Suzuki hadn't gone for water cooling, Fujji explained that it was because the air/oil-cooled engine was both narrower and lighter. A prototype liquid-cooled engine had proved this. Suzuki's attention to weight saving was a key aspect of the GSX-R and future "racer replica" designs.

AN ALUMINIUM CHASSIS

The "lowest weight" concept also led Fujji to create an aluminium chassis, both frame and rear swinging arm being constructed of this light metal. At the time, Suzuki claimed an

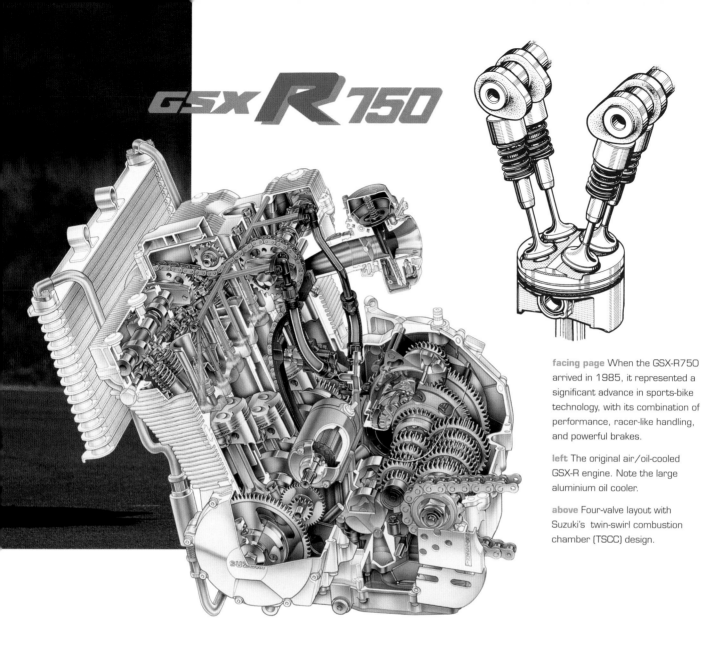

GSX R750

facing page When the GSX-R750 arrived in 1985, it represented a significant advance in sports-bike technology, with its combination of performance, racer-like handling, and powerful brakes.

left The original air/oil-cooled GSX-R engine. Note the large aluminium oil cooler.

above Four-valve layout with Suzuki's twin-swirl combustion chamber (TSCC) design.

industry first, but it was wrong – the German Ardie marque had built roadsters with alloy frames in the early 1930s. The rear suspension itself was of the monoshock variety with a vertically mounted, rising-rate shock absorber.

The exhaust was another area where weight-saving techniques were applied. The GSX-R had a four-into-one system, rather than the heavier four-into-two arrangement or four separate pipes previously favoured by other marques.

The original 1985 GSX-R750 was an uncompromising sportster, being more at home on the track than city streets. As such, it was without equal in its day. A larger-engined version, the GSX-R1100, arrived to complement, but not replace, it in 1987.

1985 SUZUKI GSX-R750F
SPECIFICATIONS

Engine Liquid-cooled, double-overhead-camshaft, 16-valve, across-the-frame, four-cylinder

Displacement 749cc

Bore and stroke 70.0 x 48.7mm

Ignition Electronic

Gearbox Six-speed, foot-change

Final drive Chain

Frame Aluminium, Deltabox

Dry weight 176.4kg (388lb)

Maximum power 104.5bhp   10,500rpm

Top speed 249km/h (155mph)

SUPER SPORT

From the late 1980s, the most popular motorcycle category has been Super Sport, which refers to 600cc, multi-cylinder machines and, to a much lesser extent, 750cc twins. These bikes combine performance aplenty with relatively low weight and lower price tags than superbikes. The class has also become popular in racing.

KAWASAKI GPZ600R

After the huge success of its GPZ900R, Kawasaki realized that it had a winning formula. It also concluded that a scaled-down, more affordable machine could, potentially, be a top seller. The result was the GPZ600R of 1985, with which the Japanese marque stole a march on the opposition and created a brand-new sector in the market – 600 Super Sport.

A landmark in modern biking history, the 600R shared many design features with its larger and older sibling, such as liquid cooling, double overhead camshafts, an across-the-frame, four-cylinder layout, and four valves per cylinder. It was far from being simply a scaled-down GPZ900R, however, since the engine, while possessing many similarities to the 900R, was actually more closely related to the contemporary air-cooled GPZ550. The 592cc (60.0 x 52.4mm bore and stroke) unit duplicated the 550's central cam chain and chain primary drive with intermediate shaft and crank-mounted alternator, but was more compact, being 40mm (1.6in) narrower.

During its first two years, the GPZ600R had the field to itself. Kawasaki's Japanese rivals struggled to compete with old-technology, air-cooled bikes.

THE OPPOSITION FIGHTS BACK

By late 1986, Yamaha was able to announce the new FZ600 (albeit still an air-cooled design), and Suzuki was working on its own 600 sportster. Honda, meanwhile, was about to drop a big spanner in the works by announcing its brand-new, liquid-cooled, 16-valve CBR600F. This went on sale in early 1987, to great press and public acclaim, and it has remained a top seller into the 21st century.

HONDA CBR600F

The CBR600F's huge success was due to Honda's ability to keep it competitive, a series of updates over the years adding to the original's great all-round abilities. Even for a company the size of Honda, the CBR's success was phenomenal.

The original CBR had fully enclosed bodywork and a steel frame. Besides the almost yearly changes to power output, the first major revision occurred for 1991 and included a new exhaust, restyled bodywork and a modified frame. By 1995, the power output had grown to 100bhp. Then, for 1999, there was virtually a new bike – including the engine – with, for the first time, an aluminium frame. In 2001, fuel injection replaced the original carburettors and power output was increased to 109bhp at 12,500rpm; in addition to the "F", a new sport model was available.

left Honda's CBR600 has been a best-seller in the middleweight super sport sector. The original CBR of 1987 had fully-enclosed bodywork and a steel frame.

above The Kawasaki GPZ600R of 1985 created a brand-new motorcycle category, the 600 Super Sport.

right In 1989, Yamaha launched the FZR600, which replaced its air-cooled FZ600. The new bike followed the company's "Genesis Concept" of inclined engine and Deltabox frame.

TOURING BIKES

For many years, touring bikes were big, heavy motorcycles with poor performance, dull handling, and less-than-perfect brakes. Modern machines, however, are entirely different.

HOW THE BREED EVOLVED
In the past, touring bikes were usually adapted from standard models, often with the addition of a sidecar. There were a few exceptions to this rule, such as Vincent's short-lived Black Knight of the mid-1950s and BMW's R60/R69S of the 1960s, but these were too expensive for most during the prolonged austere period that followed World War II.

With the arrival of the 1970s, the era of the modern superbike began, spurred on by the introduction of Honda's ground-breaking CB750 in 1969. The plethora of high-performance models that followed led to a demand for speed, but coupled to designs that could carry two people over long distances in comfort. BMW's R100RS and Moto Guzzi's SP1000 were the first of this new breed.

Today, most people in the developed world enjoy large amounts of leisure time, and for many, there is no better way to use that time than touring on two wheels. Motorcycle manufacturers have been only too eager to provide the means – in addition to the big four Japanese marques, BMW, Ducati, Triumph, Moto Guzzi, and Aprilia have all offered serious touring machines.

HONDA PAN EUROPEAN
One of the best Japanese designs has been the Honda Pan European ST. This machine is an excellent example of how to design and build the ultimate touring bike. Honda's engineering team began with a brief to create a fast, practical, and reliable mount, which it accomplished in superb fashion. The liquid-cooled V4 engine is superb, offering high torque and usable power in a smooth, unstressed fashion. This, coupled to excellent handling, powerful brakes, shaft final drive, and optional traction control system (TCS) and ABS, makes the ST one of the world's safest bikes.

BMW'S TOURING CHALLENGE
The German BMW company has always produced excellent touring models, and by the early 21st century it was the main competitor of the Japanese, with both twin- and four-cylinder models. BMW's K1200LT four-cylinder tourer was launched at the 1998 Munich motorcycle show. This giant among bikes challenged Honda's GoldWing in the heavyweight stakes at 378 kg (834 lb). Its 1,171 cc engine featured liquid cooling, double overhead camshafts, 16 valves, Bosch electronic ignition, and a six-speed gearbox.

Many, however, continued to opt for the R1100RT twin, launched in 1996, which is considerably lighter, but still boasts decent performance and a high-quality specification.

facing page With its liquid-cooled V4 engine, adjustable screen height and optional traction control, Honda's Pan European was much admired.

right The Moto Guzzi California FF 949cc, overhead-valve V-twin was offered with the choice of conventional carburettors or a fuel injection system.

below A luxury tourer: the 2006 Yamaha FJR1300A, with comprehensive weather protection and excellent integral luggage capacity.

HONDA V-FOUR

The Honda V-Four family did not get off to the best of starts. The Japanese manufacturer introduced the first mass-produced V4 motorcycle in 1982, but, as one commentator rightly said, "the 750S was met with all the enthusiasm one accords to cold porridge." Unfortunately, the machine's revolutionary engine was compromised by appalling styling, a bad riding position, and poor handling.

THE VF 750F

A year later, Honda introduced the VF 750F (Interceptor in the USA), a super sport version of the original, which performed, handled, and looked much better. It promptly outsold every other 750 that year. Unfortunately, the "F" was plagued by mechanical woes, which largely centred on camshaft wear and cam chains. By this time, Honda had extended its V4 range to include 400, 500, and 1,000cc models, the last in two guises – roadster and race replica. The public, however, had lost faith in the concept.

PUTTING THINGS RIGHT

Unused to such a humiliating response, Honda re-examined and revised the design, launching the VFR750 in 1985. In the process, it not only reinvented the V4 configuration, but also produced an all-time great. The VFR (which featured gear-driven cams) was an all-rounder, however, not a true sports bike. To fill this role, the company went a stage farther and created the RC (Racing Corporation) sports/racing machines. Subsequently, examples of these bikes gained four World

Superbike titles and proved to be Japan's main challengers to the all-conquering Ducatis during the first decade of the championship series.

RC30 AND RC45

Launched in 1988, the Honda RC30 was a genuine racer that could also serve as a street bike. Its 749.2cc, liquid-cooled, double-overhead-camshaft, 90-degree V4 engine generated 103bhp at 11,500rpm and could push the bike to 246km/h (153mph) in full road trim.

Production of the RC30 ended in 1992. It was replaced by the RC45, which was offered from 1994 to 1998. The new model copied the RC30 in many respects, including the engine size, aluminium frame, and revolutionary, single-sided rear swinging arm, which had been developed in conjunction with the French ELF company. The RC45 was lighter than the RC30, however, at 189kg (416.6lb) compared with the latter's 193kg (425.4lb).

Among the features carried over from the earlier machine were lightweight titanium con-rods, but the RC45 employed new powder-metal composite cylinder liners (the RC30 had steel items). There were also improvements to the six-speed gearbox, hydraulic clutch, and cooling system, and a revised four-into-one exhaust. The new bike was equipped with programmed fuel injection, a first on a production Honda. An inverted front fork replaced the conventional telescopic unit of the RC30, while the swinging arm was modified to accommodate a wider tyre.

facing page Carl Fogarty (1) and Arron Slight (3) with their Castrol-sponsored, works RC45 V4s, circa 1996.

above Launched in 1986, the RC30 was a genuine racer that could double as a road bike.

right The RC30's 749.2cc, liquid-cooled, double-overhead-camshaft, 90-degree V4 put out 103bhp at 11,500rpm.

below Honda also produced a series of other V4 designs, including the VF400 of 1984.

below right A feature of the RC30 (and RC45) was the French ELF-conceived, single-sided, rear swinging arm.

AERODYNAMICS

The development of the modern, high-performance motorcycle has been greatly assisted by the incredible strides made in aerodynamic streamlining over the past few decades. Initially, streamlining was employed in motorcycle speed record attempts, then was adopted for racing machines, and finally made an appearance on production roadsters. Although several motorcycle designs had some form of enclosure before World War II, proper streamlining did not become commonplace until the 1950s.

GRAND PRIX

It was in grand prix racing that the use of streamlining really flourished. At the beginning of the 1950s, machines had simple flyscreens, but by the middle of the decade, full "dustbin" fairings were the rule. Concerned that the practice was getting out of hand, the sport's governing body, the FIM, banned full streamlining – on safety grounds – from the end of the 1957 season.

From the beginning of 1958, racing machines began using the "dolphin" fairing, a much less cumbersome device than the fully streamlined "dustbin" shell of the previous year. Even today, the "dolphin" is the only form of fairing used on racing machines; it is also fitted to production street bikes. Although far from perfect, it does not suffer from problems in high side winds, which caused the "dustbin" to be banned.

THE FIRST "PRODUCTION" FAIRINGS

The first fairings to be fitted to production roadsters, as original equipment, were largely the work of British motorcycle manufacturers. In 1955, Vincent introduced the fully faired Black Knight V-twin; other notable examples, which appeared at the end of that decade, were the Royal Enfield Airflow and the Ariel Leader.

During the mid-1970s, BMW (R100RS) and Moto Guzzi (SP1000) offered faired machines, being joined, in 1979, by Ducati (Mike Hailwood Replica). For once, the Japanese had been caught off guard – even as the 1980s dawned, most of their machines were still "naked".

facing page With their low lines, kneeler racing sidecar machines display good aerodynamics.

right On the 2005 Ducati 999, the rear silencer was faired-in to improve streamlining.

below The 2003 Benelli 900 Tornado triple incorporated air ducts for the cooling and fuel injection systems.

below right The narrow frontal view of Kawasaki's GPZ600R, which was the first of the 600 super sport machines.

With the advent of the racer replica (such as the Suzuki GSX-R750) and tourer (Kawasaki GTR), however, the Japanese had soon covered up most of their bikes.

BETTER PERFORMANCE AND FUEL ECONOMY

Today, streamlining is a vital feature of motorcycle design. Not only does it improve rider comfort and machine aesthetics, but it also offers gains in performance and fuel economy. Key aims in a design team's mission are minimal frontal projected area and coefficient of drag (CdA). At around 200 km/h (125 mph), reducing the CdA of a bike by only 0.01 adds 1 km/h (0.62 mph) to the speed – equivalent to raising engine power by 1–2 bhp. At higher speeds, the increase is substantially greater. Thus, as speeds rise, aerodynamics become ever more important.

The engine of a modern streamlined motorcycle requires a good supply of cooling air to prevent overheating. Ducts to allow this to enter and escape are built into the front, sides, and rear of the fairing.

SUZUKI RGV250

During the late 1980s and early 1990s, the Suzuki RGV250 V-twin was one of the most popular motorcycles. The key to its success was its unique blend of high performance, race-like handling, and relatively low purchase price.

The origins of the design can be traced back to 1983, when Suzuki introduced its excellent RG250 parallel twin. Although this never sold in large numbers, it was the best-handling Japanese bike of its era. Not only did it pioneer the use of single-shock, rising-rate rear suspension in Suzuki's range, but it also displaced the Yamaha 250LC as the top production racer in the 250cc class.

THE RGV250 ARRIVES

The first news of the RG250's replacement came in early 1987, when details of a sensational new Suzuki 250 leaked out; it went on sale later that year. The first version to be exported was the "K", in the 1989 model year. This was followed in 1990 by the "L", which was virtually identical.

The "K" and "L" versions of the RGV250 had the same 249cc, 90-degree, V-twin engine, equipped with twin Slingshot carburettors, a newly designed induction system, a radial-flow radiator, and automatic exhaust timing control (AETC) to increase torque throughout the rev range. The result (with full road silencing) was a power output of 57 bhp at 11,000 rpm and a top speed of 209 km/h (130 mph).

ALLOY FRAME

Made from cast alloy, the RGV250's frame had been designed to minimize weight while maintaining torsional strength. It was equipped with Suzuki's "full-floater", link-type, monoshock rear suspension and triple disc brakes.

Unfortunately, the early RGV models were not particularly reliable, suffering from seized power valves, and cylinder and piston problems. Suzuki engineers tackled these shortcomings for the 1991 model year, when they virtually redesigned the RGV. The result was a much better bike with an uprated engine that featured new power valves. There was also a crescent-shaped, aluminium, rear swinging arm. This allowed both expansion-chamber exhaust outlets to discharge on the right and also permitted an amazing 58-degree banking angle.

Other developments included a coaxially mounted gear-shift lever and revised gear ratios to improve acceleration. The main frame was strengthened as well.

INVERTED FRONT FORK

The new bike featured an inverted (upside-down) front fork with a revised steering-head angle. The wheel size was 432 mm (17 in) front and rear (previously, an 457 mm/18 in rear wheel had been used), while the brakes were uprated.

Although the power output remained unchanged, a reprogrammed engine management system provided more usable power characteristics. Suzuki also offered a wide range of racing parts in "kit" form. These included close-ratio gear sets (a cassette gear cluster was employed), power-valve breathers, full racing exhausts, and a racing engine management "black box".

With kit parts and specialized tuning, 70 bhp was possible, and for several years the RGV was "top dog" in the highly competitive Super Sport 400 class (250 cc two-strokes; 400 cc four-strokes).

1989 SUZUKI RGV250K
SPECIFICATIONS

Engine Liquid-cooled, reed-valve, two-stroke, 90-degree, V-twin

Displacement 249 cc

Bore and stroke 54.0 x 54.0 mm

Ignition Electronic

Gearbox Six-speed, foot-change

Final drive Chain

Frame Aluminium, Deltabox

Dry weight 128 kg (282 lb)

Maximum power 57 bhp at 11,000 rpm

Top speed 209 km/h (130 mph)

HONDA FIREBLADE

Two motorcycles dominated the sports-bike scene during the 1990s: the fabulous Ducati 916 and Honda's equally brilliant CBR900RR FireBlade.

GETTING IN FIRST

Arriving first, in 1992, the FireBlade immediately put its rivals at a disadvantage with its combination of very light weight and stunning power. Honda's engineering team, led by Tadeo Baba, had conceived an 893cc, liquid-cooled, double-overhead-camshaft, 16-valve, across-the-frame four, which, in its original "N" guise, developed 122bhp at 10,500rpm, giving a top speed of 266km/h (165mph). With its low 186.5kg (411lb) weight, the machine's power-to-weight ratio was unmatched.

Only the sub-900cc engine size was questioned, but this was rectified when increased to 918cc for 1996, then 929cc for 2000, and 998cc for 2004.

QUEST FOR MINIMUM WEIGHT

The FireBlade's impressive technical specification, for a series-production bike, helps to explain why the power-to-weight ratio was so good. Baba and his team had set out to create a no-compromise sportster in the quest for minimum weight.

Flat-topped, ultra-lightweight slipper pistons gave a high degree of rigidity, while equally lightweight con-rods and crankshaft minimized inertia and provided an extremely crisp engine response. Another weight-saving measure was the use of magnesium instead of aluminium for the cylinder head (from the 1994 model year).

The FireBlade's 16 valves were operated directly by twin camshafts, resulting in a compact, low-friction valve-train. This required adjustment only every 25,700km (16,000 miles).

CARBURETTORS TO FUEL INJECTION

Originally, a bank of four 38mm, flat-slide carburettors fed the engine with fuel through a single-port induction system. But from 2000, an electronically controlled, PGM-F1 fuel injection system did this task, giving a more precise throttle response.

A cartridge-type, liquid-cooled oil cooler was fitted to keep lubricant temperatures under control for stable output and maximum engine life.

facing page From the 2000 model year, the FireBlade was equipped with fuel injection in place of carburettors; also new that year was a displacement of 929 cc.

left The first FireBlade model appeared in 1992.

below The original FireBlade had an engine size of 893 cc and developed 122 bhp at 10,500 rpm. It was the work of Japanese designer Tadeo Baba.

ALUMINIUM FRAME

The chassis played a vital role in the FireBlade's success and was subjected to a steady evolutionary process. From the very beginning, in 1992, it was made of aluminium. By the 1997 model, the frame was based on two main spars of extruded aluminium, weighing a mere 10.5 kg (23 lb). Featuring a semi-floating rubber engine mount to combat vibration, it supported an aluminium, rear swinging arm that had been designed with the aid of computer analysis and rigorous track testing.

For 2000, radical changes were made to the FireBlade. In addition to the fuel injection system, there was the larger (929 cc) engine size, a longer rear swinging arm, larger front brake discs, and inverted front fork (a first for Honda's big-bore sports models). To enable the machine to compete in the revised World Superbike series, Honda increased the engine size to 998 cc for 2004.

2000 HONDA FIREBLADE
SPECIFICATIONS

Engine Liquid-cooled, double-overhead-camshaft, 16-valve, across-the-frame, four-cylinder

Displacement 929 cc

Bore and stroke 74.0 x 54.0 mm

Ignition Digital with integrated fuel injection

Gearbox Six-speed, foot-change

Final drive Chain

Frame Aluminium, triple box spars, multi-point diamond configuration

Dry weight 170 kg (373.7 lb)

Maximum power 152 bhp at 11,000 rpm

Top speed 273.5 km/h (170 mph)

9 LIFESTYL

E AND
LEISURE

LIFESTYLE AND LEISURE

By the beginning of the 21st century, the motorcycle was no longer seen primarily as a means of transport, rather it had become an aspect of lifestyle and leisure. The latest designs and innovations would have seemed as remote as the prospect of space travel 100 years ago, at the dawn of the internal-combustion engine and the first motorcycles.

Not only has the whole purpose of the motorcycle changed, but also the technology behind it. Today, like cars, motorcycles are no longer constructed in a manner that allows them to be serviced by the home mechanic. Instead, complex computerized systems govern many of the operations. This is due to both technical advances and changes in the industry, where dealer servicing has become the norm, with attendant commercial benefits. The situation also goes some way to explaining why classic bikes have become so popular – as a rule, a motorcycle manufactured before the end of the 1970s can be rebuilt by a knowledgeable enthusiast with only limited workshop facilities. With the advent of such features as solid-state electrics, digital ignition, and fuel injection, however, even a conventional workshop manual is no longer of use; the skilled mechanic has been replaced by the computer.

Moreover, the industry makes far more specialized machines. In the past, manufacturers built a single model that could cope with many tasks, such as commuting, touring, clubman's racing, and even off-road forays. That is no longer the case.

Today, the motorcycle market has defined sectors: sports bikes (many of which require little modification to go racing), tourers, enduro-styled on/off-roaders, custom

cruisers, retro-styled machines, motocrossers, scooters, trials bikes, and superbikes. Whatever your requirements, there is a bike designed specifically to meet that need.

It would be true to say that no bad motorcycles are manufactured today; something that certainly could not have been said even 20 years ago.

The renowned American manufacturer Harley-Davidson really started the "lifestyle" culture in motorcycling. Other marques have followed the company's lead, with varying degrees of success.

But what of the future? Well, safety is certain to become more of an issue than it is today. Already, experiments have been carried out with leg protectors and even air bags on motorcycles. BMW attempted to set an example with a self-imposed 100bhp engine power limit in

the 1980s and 1990s but, since then, has been forced to join the power race. As engine sizes and power outputs creep ever higher, it may only be a matter of time before governments step in and call a halt through legislation. Perhaps smaller-capacity, more fuel-efficient motorcycles will make a return sometime in the near future.

These and other important issues that affect motorcycle design are already being studied by manufacturers' engineering teams around the world.

Finally, it is interesting to speculate on which nation will be the next major motorcycle power – China maybe?

CUSTOM BIKES

The factory-built custom bike owes its existence, primarily, to the 1976 film *Easy Rider*. Featuring a pair of "chopped" Harley-Davidsons, breathtaking American landscapes and contemporary sound track, the movie caused a whole generation to become converts to a new motorcycle sub-culture, almost overnight. It encouraged fantasies of the custom-cruiser life, wherein the sun always shone, the air was clean, and the highway stretched, arrow-straight, as far as the eye could see.

At first, many enthusiasts "chopped" existing motorcycles. A whole new cottage industry was established (in similar fashion to the earlier café racer cult) to meet their needs, producing everything from "King and Queen" seats through "peanut" tanks to complete bikes, and even trikes.

Harley-Davidson has benefited substantially from building bikes that reflect the custom-cruiser lifestyle image, as have Honda, Kawasaki, Yamaha, and Suzuki. All four Japanese marques have produced myriad Harley-like V-twins, ranging from the small to the truly massive. BMW has also muscled into the act with its R1200C (Custom) flat-twin cruiser.

YAMAHA XV535 VIRAGO

Yamaha's XV535 Virago was a brilliant design. For years, the Japanese manufacturers had tried everything to produce a top-selling custom bike, but with little success. The XV535 changed all that. Introduced at the end of the 1980s, it remained a popular bike in the early years of the 21st century. Developing 47 bhp, its single-overhead-camshaft, V-twin engine

could propel the machine to a genuine 160 km/h (100 mph). With a choice of flat or pull-back handlebars, shaft final drive, front disc brake (rear drum), a dry weight of 182 kg (401 lb), it was very easy to ride. As a result, it was responsible for introducing many newcomers to biking – and for encouraging more than a few ex-motorcyclists to return.

HONDA VTX1800

The much more recent Honda VTX1800 was right at the other end of the custom-cruiser spectrum. This monster was powered by a liquid-cooled, single-overhead-camshaft, six-valve, 52-degree V-twin. It displayed a combination of technical innovation and traditional approach in its design. For example, at the front, it had an inverted fork and six brake caliper pistons, while at the rear were two fully enclosed, chromed shocks. Weighing in at a hefty 321 kg (708 lb), the VTX produced 95 bhp at 5,000 rpm.

BMW R1200C

As might be expected, the R1200C had BMW's characteristic flat-twin engine. At the time of its launch, in 1998, the 1,170 cc (101.0 x 73.0 mm bore and stroke), four-valves-per-cylinder, high-camshaft unit was the largest of its type produced by the German marque. The machine not only employed the latest Motronic engine management system, but was also the first custom cruiser to feature exhaust emission control by closed-loop, three-way catalytic converter as standard.

left Honda's VTX1800 has masses of chrome-plated components, including the engine outer covers, the exhaust and air cleaner. It is a massive machine, weighing 321 kg (708 lb).

below The 2005 VTX1800 has an advanced specification, which includes a liquid-cooled, six-valve, 52-degree, V-twin engine.

facing page A pair of Kawasaki Eliminator 600s. Excellent performance came from a detuned ZZR four-cylinder engine and relatively light weight.

below BMW's R1200C was the first custom model from the famous German marque. It was a high-quality product with a 1,170cc, four-valves-per-cylinder, high-camshaft, flat-twin power unit.

RETRO BIKES

The retro bike combines "yesteryear" looks (without the all-enveloping plastic bodywork of most modern machines), the latest technology, and a "street fighter" image.

BIRTH OF THE CONCEPT

Today, retro bikes occupy an important sector of the market. This style of machine was conceived during the late 1980s, when the Italian Moto Guzzi company launched the nostalgic 1000S. This newcomer differed from its contemporaries in that it clearly aped the classic 750S3 model of the early 1970s, but with an updated specification.

Perhaps without realizing it, the long-established Italian marque had created an entirely new breed of motorcycle, the retro model. The dictionary defines the word "retrospect" as

"when you look back", which is exactly what Guzzi had done. "Retrospect" hardly trips off the tongue, however, so the industry's marketing men shortened it to "retro". Since then, many retro machines have been performance orientated, not simply rehashes of old favourites, and usually the engineering has been bang up to date.

THE ZEPHYR

The first Japanese company to exploit the retro concept was Kawasaki, with its Zephyr series, initially a Japanese-market-only 400, which was introduced in 1989. A larger-engined model arrived in 1990, basically the same machine bored and stroked to 553cc; later still came the 750 and 1,000 Zephyrs. All copied the high-performance, air-cooled Kawasaki

facing page A 1999 Laverda 750 Strike. It's retro styling hid an advanced specification, which included a double-overhead-camshaft, parallel twin engine, six speeds, inverted front fork and aluminium, twin-spar frame.

left Kawasaki's Zephyr 550 was introduced in Japan and the USA in 1990, and in Europe in 1991. It is largely credited with starting the retro fashion.

below Two of the most popular retros of 2005, the Triumph Bonneville (left) and Kawasaki W650. Both had parallel twin engines, the British bike's being an overhead-valve unit, while the Japanese machine had an overhead camshaft.

fours of the previous two decades, and specifically the 1000R Eddie Lawson Replica, "arguably Kawasaki's ultimate hot rod", as *Cycle* magazine reported in its 1990 road test of the 550 Zephyr. Later in the decade, Kawasaki ditched the Zephyr series in favour of the excellent "pure classic" W650 twin. This looked like a 1960 Triumph Bonneville, but came with an eight-valve, overhead-cam engine, electric starter, digital ignition, five-speed gearbox, and front disc brake.

SUZUKI BANDIT

Although Suzuki was the last to enter the retro market, it was certainly the most successful, with its outstanding 600 Bandit (1995) and 1,200 Bandit (1996). Both became top sellers – on merit. Compared with the earlier Japanese retros, the Bandit models scored by having modern, single-shock rear suspension, while their rivals retained the old-fashioned, "classic" twin-shock arrangement. The Suzukis also gained by being much lighter in weight.

THE EUROPEAN RESPONSE

In addition to the reborn British Triumph marque, which launched its Bonneville retro in 2000, BMW, Laverda, and Ducati have all produced their own interpretations of the theme. Like the Suzuki, the Ducati Monster proved a hit. Designed by Miguel Angel Galluzzi, it was introduced at the Cologne motorcycle show in late 1992. Subsequently, it was produced in 600, 750, 900, and 1,000 engine sizes, in both two- and four-valves-per-cylinder guises.

DUCATI

For almost three decades, the design talents of Fabio Taglioni had guided Ducati. He had been responsible for creating the singles and V-twins that had established the Italian company as a household name among motorcyclists.

FOUR-VALVE TECHNOLOGY

Taglioni, however, had chosen largely to ignore the development potential of four-valve technology. This was left to his successor, Massimo Bordi.

The first of Bordi's four-valves-per-cylinder models made its debut in 1986 at the French Bol d'Or 24-hour endurance event. Although it retired after running for eight hours, it had shown great potential.

The Cagiva takeover of Ducati in 1985 had provided the funding for the project, and the first production model, the 851, arrived in 1988. In the same year, Marco Lucchinelli used the bike to win the first ever World Superbike race at the British Donington Park circuit.

The standard 851 had bore and stroke dimensions of 92.0x64.0mm, but the bore in Lucchinelli's bike measured 94.0mm, giving 888cc.

Ducati won its initial World Superbike title in 1990 (the first of a series) with Raymond Roche. The company built the 851 and later 888 models for sale as both racers and street bikes, one of them being the limited-edition SP (Sport Production). The 851 was discontinued in 1993, and in 1994, the Corsa (Racing) model's engine size was increased to 926cc.

THE 916

The really big news for 1994, however, was the arrival of the ground-breaking 916. This machine's engine size matched its code number and had been achieved by increasing the stroke of the 888 from 64 to 66mm.

The remainder of the 916, though, set it apart from earlier Ducatis. This was the work of Massimo Tamburini, who has been described as one of the greatest motorcycle designers of the late 20th century – and with good cause.

To many enthusiasts around the world, the 916 was not simply the latest superbike, but the best there had ever been. It set new standards of performance, handling, and braking, but also had style and charisma. The choice of a Ferrari-red paint scheme added to the overall effect, as did the twin-head-lamp fairing, high-level silencers, single-sided, rear swinging arm, and magical, deep-throated, booming exhaust note.

996, 998, AND 999

The 916 changed the face of sports bikes in a way that only the original Suzuki GSX-R750 had managed a decade earlier.

To further improve its chances, Ducati employed a 996cc (98.0x66.0mm bore and stroke) engine in its World Superbike racers – and, from 1999, in its series-production roadsters. Later, this was enlarged even more, to 998cc, before, in 2004, being replaced by the Pierre Terblance-styled 999. For many enthusiasts, however, the Tamburini styling of the 916/996 era was best.

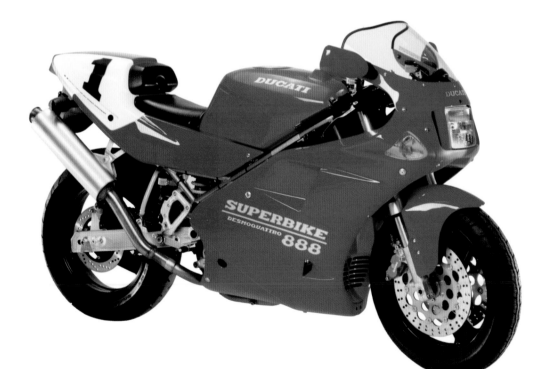

left Introduced in 1987, the Ducati 851 V-twin was the first of the marque's bikes to feature four valves per cylinder, liquid cooling, and fuel injection. The 888 followed; this is a 1993 SP5 model.

facing page:
top The Ducati 916 replaced the 888 and had all-new styling.

bottom Ducati dominated the World Superbike series during the 1990s with the 916 and, later, the 996.

HARLEY-DAVIDSON: THE MODERN ERA

Harley-Davidson motorcycles – like Porsches in the car world – are shining examples of how a primitive design can be refined until it is comparable with many, much more technically advanced machines. Also, like the renowned German cars, Harley-Davidsons are considered icons.

For some of its 100-plus-year history, however, Harley-Davidson didn't enjoy the reverence it receives today. During that period (1968–81), the marque was under the control of the American Machine & Foundry Company (AMF).

THE AMF TAKEOVER

Formed in 1903, Harley-Davidson remained a family dominated business until it became a public-owned company in 1965. In 1968, it was taken over by AMF, which, although willing to invest large sums of money, unfortunately didn't understand the motorcycle business. This resulted in poor-quality products and lacklustre marketing. Not surprisingly, the company's market share fell while losses rose. Finally, during 1981–82, a management buy-out took place, which brought a revival of the famous marque's fortunes.

"NEW NOSTALGIA"

One of the main facets of the revival plan would become known as "new nostalgia", the machines' somewhat "dated" looks being emphasized as a positive asset. This trend-setting idea was inspired by Willie G. Davidson and led to Harley-Davidson's V-twin street bikes being promoted as the motorcycles to be

seen on. This was helped by the fact that many famous personalities – film stars, sportsmen, and even politicians – became Harley owners.

In a clever marketing ploy, the company created a whole family of nostalgically styled V-twins, among them the 883 Sportster, XL1200 Sportster Custom, FXDL Road King (based on the long-running Electra Glide), and the FLSTCN Heritage Softail Special. All had variations of the legendary 45-degree, V-twin engine, which had been given a major facelift as the V2 Evolution in the mid-1980s.

This was a classic Harley-Davidson update, not a back-to-the-drawing-board (or computer, as it is now) project. Although the old and the new engines look similar externally, the latter was improved in almost every aspect. For example, the alloy cylinder barrels with iron liners were far more efficient at removing excess heat. The entire engine was lighter, more powerful and easier to maintain.

THE ACCESSORY MARKET

Harley's biggest success, however, has been its ability to encourage many owners to personalize their bikes with after-market accessories. In fact, the commercial success enjoyed by the company since the mid-1980s has been built on this strategy. This is where the real profits have been generated and why Harley-Davidson has managed to grow year-on-year in such a storming fashion, with bikes that are based on what is, essentially, an outdated design.

left Designed as the most affordable of the modern Harley-Davidsons, the 883 model allowed owners to benefit from the aura surrounding the famous American marque.

facing page:

top In the early 21st century, Harley-Davidson introduced the liquid-cooled, 1,130cc, VRSCA V-Rod, with 115bhp and a top speed of 216km/h (134mph).

bottom Buell is the sports-bike division of Harley-Davidson, former racer Erik Buell being responsible for the lightweight, monoshock chassis design.

OFF-ROAD RACERS

Off-road racing has boomed in recent years. It is an exciting sport for riders and spectators alike; with the steady growth in leisure time and living standards, many have been encouraged to take it up.

Although European marques, among them KTM and Husqvarna, are still involved, the Japanese manufacturers have made much of the recent running. Honda, for example, offers a vast range of off-road machinery, ranging from the CRF50F (with 49.4cc, single-overhead-camshaft, two-valve horizontal single) to range-topping models such as the XR650R (enduro), CRF 450x (motocross), and the FMX 650 (which Honda classifies as "Adventure", but is actually a Super Motad racer).

AN ALUMINIUM CHASSIS
In 1997, Honda took motocross technology into a new era, at least as far as production models were concerned, when it introduced its CR250. This had a trend-setting, aluminium, twin-spar frame, which set it apart from the competition.

In preceding years, there had been a host of important developments. These included advances in two-stroke technology (such as reed and power valves, liquid cooling and map-type, digital electronic ignition), five- and six-speed gearboxes, aluminium swinging arms, rising-rate, single-shock rear suspension, front and rear disc brakes, inverted front forks, and plastic bodywork.

In addition to its aluminium frame, the CR250 had a 249cc (66.4x72.0mm bore and stroke), two-stroke, single-cylinder engine with a composite racing valve (CRV), an all-new power-jet-control (PJC) carburettor and a host of other improvements compared with earlier versions.

By 2005, the CR250 generated 58bhp at 8,000rpm. It had been joined by the CRF250 with a 249.4cc, liquid-cooled, single-overhead-camshaft, four-valve, four-stroke engine.

FOUR-STROKES
At the beginning of the 21st century, the previously dominant two-stroke was supplemented by a new breed of high-performance, high-tech, four-stroke, off-road racer. Honda offered 50, 70, 100, 250, 450, and 650 models, the last for enduro and Super Motad events.

New for 2005 was the road-legal FMX650, with a 644cc, air-cooled, single-overhead-camshaft engine that produced 37bhp at 5,750rpm. The equally new CRF450x was a genuine motocross racer, displacing 449cc, with liquid cooling, four valves, and single overhead camshaft. Honda claimed a power output of 48.5bhp at 7,500rpm.

The most amazing machine in the 2005 Honda off-road line-up, however, was the CR125R. Its liquid-cooled, single-cylinder, two-stroke engine put out an amazing 42bhp at a screaming 11,500rpm.

facing page Part of an impressive line-up of high-tech dirt bikes from Honda for 2005. Left to right: TRM Super Moto, FMX650 and CRF450R.

above The FMX650 644cc, air-cooled, single-overhead-camshaft, single-cylinder engine with dual exhaust ports produced 37bhp at 5,750rpm.

left Motocross champion and Kawasaki works rider Kurt Nichols, with a KM5000 dirt racer powered by a liquid-cooled, reed-valve, two-stroke motor.

V-TWINS

V-twin engines have been around since the dawn of motor-cycling, but the latest examples are very sophisticated, and most offer breathtaking performance. In fact, modern V-twins are so good that, in recent years, bikes such as the Aprilia RSV Mille, Honda VTR 1000 SP-2 and Firestorm, Suzuki TL1000, French Voxon, Ducati 999, and Yamaha MT-01 are comparable to the best of the multi-cylinder machines.

THE REBIRTH

Many of the classic V-twin bikes, such as the Brough Superior and Vincent, are the top icons of their eras. Production of the Brough ended in 1939, of course, and the Vincent in 1955. In the old days, the angle between the cylinders could be anything from almost zero (Matchless Silver Arrow) to 120 degrees (as was the case in the Moto Guzzi 500 racer ridden to victory by Stanley Woods in the 1935 Senior TT).

Ducati's Fabio Taglioni laid the groundwork for the modern V-twin, certainly as regards high-performance sports models. His successor, Massimo Bordi, took things a stage farther with liquid cooling, the use of four valves per cylinder and, most importantly, an integrated electronic fuel injection and ignition system developed in conjunction with Weber-Marelli.

THE APRILIA

The Aprilia RSV differed considerably from Ducati's V-series. For a start, it had a narrower angle between the cylinders – 60 degrees instead of 90. At the time of its launch in 1998, Georgio del Ton, Aprilia's head of engine development, said, "With everyone else jumping on the 90-degree bandwagon (Honda and Suzuki, for example), we felt it necessary to establish our own identity. We compared both layouts carefully. The 90-degree engine has many advantages, but we felt there were too many compromises with its design when building a lightweight, compact, sporting motorcycle. For example, we were able to keep the RSV's wheelbase in check without resort-ing to trick alternative rear shocks and side-mounted radiators. It also allowed us to keep fuel injection components close together, which is critical to allowing straight, short inlet tracts for the cylinder heads." Another departure from the Ducati formula was the use of dry- rather than wet-sump lubrication.

YAMAHA

During the early 1980s, Yamaha marketed the 981cc TR1, which employed a 75-degree, overhead-camshaft V-twin, and had single-shock rear suspension. This never sold well, however, and eventually it was dropped from the range.

Then, in 2005, the Japanese marque launched the incred-ible 1,670cc MT-01. Featuring a big-bore, long-stroke engine with stunning high-torque thrust, it bristled with innovations. Among these were digital fuel injection; forged pistons and ceramic-coated cylinders; a fully adjustable, 43mm, inverted front fork; an alloy frame; and an R1-derived, truss-style, rear swinging arm.

In the MT-01, with its jaw-dropping performance, the long-running V-twin concept had been more than simply modernized – it had been redrawn completely.

facing page:

left Cagiva's Gran Canyon had an air-cooled Ducati 900SS engine (shown) initially, and from 2000, a Suzuki TL1000 unit. Both were 90-degree V-twins.

right The Aprilia RSV Mille, equipped with a liquid-cooled, double-overhead-camshaft, 60-degree, V-twin engine. The camshaft drive combined chains and gears.

top left and right Features of the 2005 Yamaha MT-01: matching, white-faced instruments (left); and futuristic headlamp design.

above The MT-01 was powered by a 1,670cc, overhead-valve, V-twin engine with four valves per cylinder, fuel injection, and dry-sump lubrication.

left The Voxon was yet another new breed of V-twin, offering excellent performance, plenty of torque, and unique styling.

COLOUR CODING

During the early decades of motorcycling, bikes were usually finished in drab colour schemes – black and dark greens, reds or blues. Then, toward the end of the 1950s, the Italian manufacturers began to give their sporting lightweights eye-catching finishes. Ducati (and others), for example, sold machines with gold frames and metallic cherry red tanks, side panels and mudguards.

COLOURED LEATHERS

In the mid-1960s, racers began wearing coloured leathers – before, they had been universally black. This fashion had its roots in the United States, but was soon taken up in Europe, where the Englishman Rod Scivyer gained more fame by wearing a set of white leathers than he did by winning the 125 cc British Road Racing Championship title.

In the 1970s, with the dawn of the superbike era, colour coding began to take hold as riders, particularly in the USA and Germany, began matching their riding gear to their bikes.

THE MANUFACTURERS CATCH ON

At first, coloured leathers, boots, and gloves had to be made individually by specialists. As the 1970s gave way to the 1980s, however, motorcycle manufacturers such as Honda, Kawasaki, Suzuki, and Yamaha began to appreciate that they could benefit by meeting this new demand. Not only was there

money to be made from the sale of such clothing, but also free advertising to be gained from the addition of their logos.

Kawasaki began finishing its bikes in a shocking bright green livery, which proved a great success. As a result, its factory racers, motocrossers, and road bikes established the marque's famous "Mean Green" nickname.

MOBILE BILLBOARDS

As riders and bikes became increasingly colourful, so they became adorned with sponsors' logos. Prior to the 1970s, racing machines and their riders did not carry any kind of corporate badging. By the late 1970s, however, they had begun to resemble advertising billboards. Today, this is accepted as the norm, prompted by the television coverage of such events as Moto GP and World Superbike racing.

IMITATION

Many of today's riders of road bikes are keen to exploit this "branding", not only to re-create the image that their racing heroes portray on the track, but also to share in the brand image, whether it be Honda, Suzuki, Yamaha, Kawasaki, Ducati, Triumph, or the like. This in turn has provided a ready market for identical colourful bike liveries, riding gear, and associated items. Manufacturers are more than happy to meet this demand.

facing page Colour coding applies not only to the motorcycle, but also to the modern rider's gear.

right Moto GP rider Alex Barros (4) with his factory-backed Honda during the 2005 season, displaying his distinctive yellow and blue colours.

below World Champion Valentino Rossi (46, Yamaha) leads Ducati's Loris Capirossi. The Japanese marque's red and white scheme makes a striking contrast to the Italian bike's all-red livery.

TRIUMPH REBORN

One of the real success stories of recent times is Triumph's revival under its new owner, John Bloor. The famous British marque's range of three- and four-cylinder bikes has successfully challenged the best that Japan has to offer in the showrooms – and on the road.

After buying the rights to the Triumph name in the early 1980s, Bloor brought the project to life at the Cologne motorcycle show in September 1990. By 1996, a new Triumph was rolling off the production line every six minutes, giving an annual production figure of 15,000 machines. Since then, output has risen dramatically. Today, Triumphs are exported to such diverse destinations as Australia, Brazil, Kuwait, Mexico, Russia, and Taiwan, as well as North America and Europe.

The 2005 Triumph range included machines labelled Cruisers, Urban Sports, and Modern Classics.

THE AMAZING ROCKET III

A few years ago, any kind of cruiser with the sheer size and performance of the Rocket III would have been an impossible dream. Triumph, however, made that dream an awesome reality.

The machine's three inline, fuel-injected cylinders displaced a massive 2,294cc, and produced 140bhp and 219kg/m (147lb/ft) of torque. This output was fed to a giant, 240-section tyre through a five-speed gearbox and shaft final drive. Keeping all that power under control was an extremely strong, steel frame, inverted front fork, twin rear shocks, and triple disc brakes with four-piston calipers from the marque's 955i sportster. Other engine features included double overhead camshafts and digital ignition. All things considered, its weight of 320kg (704lb) was less than might have been expected.

SPEED TRIPLE

The Speed Triple was introduced in the mid-1990s and, subsequently, was updated considerably. It was given a new 1,050cc, across-the-frame, three-cylinder engine, which provided plenty of power (128bhp) and excellent torque. Along with the bigger engine came a new gearbox with improved

T509 *SPEED triple*

facing page Under the owner-ship of John Bloor, Triumph re-established itself as a leading motorcycle marque during the late 1990s and early 21st century. Here, Francis Williamson crests the famous Mountain at Cadwell Park on a T595 triple, circa 1998.

right A 2005 Triumph T509 Speed Triple.

below With its 2,294 cc, fuel-injected, three-cylinder engine, Triumph's Rocket III is "King of the Cruisers".

below right Triumph's styling studio came up with this striking rendition of the British Union Jack flag.

action. There was also a new aluminium frame, fully adjustable, inverted front fork, and a remote-reservoir, rear monoshock. The "street fighter" style that made the Speed Triple so popular, however, was retained.

THRUXTON 900

Triumph launched its new "classic" Bonneville 790 cc, air-cooled, vertical twin in 2001. Four years later, it created the Thruxton 900, with a tuned and bored (from 86 to 90 mm) version of the existing Bonneville motor. The newcomer was particularly successful in recapturing the spirit and style of the 1960s café racer era. With clip-on handlebars, rear-set footrests, twin upswept megaphone silencers, chrome head-lamp, and evocative seat hump, it couldn't help but impress potential buyers.

In the early 21st century, with a range that encompassed super sports machines, tourers, trail bikes, and cruisers, the reborn Triumph marque was back among the leaders in motorcycle design.

YAMAHA R1

When it appeared at the end of 1997, the Yamaha R1 caused a sensation. The 1,000cc bike, in full road trim, weighed an amazingly low 177kg (390lb) – less than the majority of contemporary super sport 600s!

The R1 project began in 1995, when Yamaha engineer Kunihito Miwa set about designing and building the fastest, lightest, and best-handling sports bike possible. The result was considerably smaller and lighter than its predecessor, the YZF 1000R (Thunderace).

A COMPACT ENGINE

The R1 featured a compact version of Yamaha's familiar 20-valve, four-cylinder, across-the-frame engine. Displacing 998cc (74.0x58.0mm bore and stroke), it employed a one-piece cylinder and crankcase assembly. A sophisticated exhaust ultimate power valve (EXUP) system took into account engine

speed, throttle position, and even the rate at which the throttle was opened. It also provided additional mid-range torque.

The R1's engine sat well forward in the aluminium, twin-beam, Deltabox frame, allowing a longer rear swinging arm (also aluminium) to improve stability.

MULTI-ADJUSTABLE SUSPENSION

The suspension was multi-adjustable, the inverted front fork featuring extra rebound at the end of its travel to help keep the front wheel down under hard acceleration. Miwa and his engineers also paid particular attention to the brakes and wheels (150mm/6in wide at the rear with a 190-section tyre).

Although largely conceived as a roadster, in many ways, the R1 had been designed like a race bike, particularly in respect of its "minimum weight" theme. For example, the four-piston, front brake calipers had been lightened to reduce

facing page Just as Suzuki's GSX-R changed 750cc sports-bike design in 1985, so did Yamaha's 1,000cc R1 when it appeared in late 1997, with its ultra-lightweight chassis and awesome performance.

left Designer Kunihito Miwa and his team gave the R1 some nice touches, including these footrest controls…

below …and streamlined, car-type headlamps.

unsprung weight, while the three-spoke alloy wheels were hollow. Cutting weight, Miwa reasoned, helped improve suspension performance.

ENGINE TRACTABILITY

Another aspect of the engine (common to all big-bore, five-valves-per-cylinder Yamaha engines) was its tractability. In standard, untuned form, the bike could be ridden smoothly at 32 km/h (20 mph) in top gear and, with little or no hesitation, accelerated all the way to the 11,500 rpm red line.

Miwa updated the R1 for 2000, notably with a new fairing to improve aerodynamic efficiency and reduce rider fatigue. There was also a new titanium silencer.

Suzuki's answer to the R1 was the GSX-R1000; Kawasaki introduced the ZX-10. This led to an ongoing battle between the three manufacturers in the showroom and on the track.

1998 YAMAHA R1 SPECIFICATIONS

Engine Liquid-cooled, double-overhead-camshaft, 20-valve, across-the-frame, four-cylinder

Displacement 998cc

Bore and stroke 74.0x58.0mm

Ignition Electronic

Gearbox Six-speed, foot-change

Final drive Chain

Frame Aluminium, twin-beam, Deltabox

Dry weight 177kg (390lb)

Maximum power 160bhp at 11,500rpm

Top speed 282km/h (175mph)

MV AGUSTA F4

Hailed by many as the ultimate modern motorcycle, the MV Agusta F4 could easily have been a Cagiva or Ducati – or even a Ferrari. The fact that it was an MV is a fitting tribute to one of the strangest of all bike projects.

BIRTH OF AN IDEA
The original concept for the machine came from conversations between Claudio Castiglioni, Cagiva's boss, and his chief designer, Massimo Tamburini. Piero Ferrari, son of Enzo Ferrari, also contributed.

The first rumours concerning a brand-new, Cagiva-inspired sports bike began to circulate at the beginning of the 1990s. Then, at the launch of the Ferrari 465 GT car, a photograph of the new motorcycle engine was shown by mistake. As a result, Castiglioni was forced to confirm that it was being developed in conjunction with Ferrari, and that he and Piero Ferrari had actually tested the prototype machine.

DEVELOPMENT GLITCHES
Cagiva suffered financial problems during the mid-1990s, which slowed development at the company's secret hilltop research centre in San Marino, but in the spring of 1998, the first example of the long-awaited machine appeared. From the start, it was badged as an MV Agusta. (Cagiva had purchased the famous marque's name some years earlier, and, having sold Ducati to the American Texas Pacific Group, it could not have used that name.)

The F4's styling was superior to that of every other series-production sports bike in the world, while it bristled with innovative technical features.

TECHNICAL DETAILS
The 749.8cc (73.8x43.8mm bore and stroke), liquid-cooled, four-valves-per-cylinder, double-overhead-camshaft, across-the-frame, four-cylinder engine was equipped with a radical, radial-

facing page The four-cylinder MV Agusta F4 series was launched in 1998 with a 750-class sportster. A larger-engined version, the F4 1000, arrived in 2004. Shown is the limited-edition 1000 Ago.

above The F4's styling was the work of Massimo Tamburini, who created the Ducati 916.

left The Brutale appeared in 2002. This is the 2005 America with its distinctive red, blue, and white finish.

valve cylinder head, and the very latest Weber-Marelli electronic fuel injection and ignition systems. It was combined with a removable (cassette-type), six-speed gearbox.

The F4's frame was a combination of steel and aluminium components, while its steering-head angle was fully adjustable. At the front was a specially built, inverted Showa fork assembly with an Öhlins hydraulic steering damper mounted across the frame. The exhaust outlets were beneath the seat, this being a Tamburini styling trademark.

Tamburini also conceived a particularly innovative front light arrangement. The twin stacked polyellipsoidal units gave the F4 a unique appearance, while allowing a very narrow fairing for maximum aerodynamic advantage.

A naked version, stripped of its bodywork, the Brutale, was introduced in 2002. By 2005 there was a larger-engined F4 1000 sportster, plus the distinctively customized America with its eye-catching red, blue, and white paint job.

2000 MV AGUSTA F4 750
SPECIFICATIONS

Engine Liquid-cooled, double-overhead-camshaft, 16-valve, across-the-frame, four-cylinder

Displacement 749cc

Bore and stroke 73.8x43.8mm

Ignition Electronic, integrated with fuel injection

Gearbox Six-speed, foot-change, cassette type

Final drive Chain

Frame Steel and aluminium, fully-adjustable steering head

Dry weight 180kg (397lb)

Maximum power 137bhp at 12,600rpm

Top speed 282km/h (175mph)

GLOSSARY

Aerodynamics A science that focuses on the movement of solid forms through the air.

Air cooling A method of removing heat from engine cylinder barrels and heads by the natural flow of air over them. The components are finned to aid the process.

AIV Automatic inlet valve. Activated by engine suction; forerunner of the mechanically operated valve.

Belt drive A leather or fabric belt running from the engine or gearbox to the rear wheel.

Bhp Brake horsepower. A measure of an engine's power output.

Bore and stroke ratio The ratio of an engine's cylinder diameter to its piston stroke.

Boxer An engine with horizontal cylinders.

Caliper A clamping device containing hydraulically operated pistons; part of a disc brake.

Cam A device for opening and closing a valve.

Camshaft The mounting shaft for a cam; can be in low, high or overhead position.

Carburettor Produces the air/fuel mixture for the engine.

Chain drive Primary form of drive from engine to gearbox, and secondary (final) from gearbox to rear wheel.

Combustion chamber Area where the fuel/air mixture is compressed and ignited, between the piston and cylinder head.

Connecting rod Assembly that joins the piston to the crankshaft; also con-rod.

Compression ratio The amount by which the fuel/air mixture is compressed by the piston in the combustion chamber.

Crankcase The casing enclosing the crankshaft (and gearbox in a unit-construction engine).

Crankshaft The shaft that converts the vertical motion of the piston into a rotary movement.

Cylinder Contains the piston and is capped by the cylinder head. Upper portion forms the combustion chamber where the fuel/air mixture is compressed and burned to provide power.

Cylinder head Caps the top of the cylinder. In a four-stroke engine, it usually houses the valves and, in some cases, one or two camshafts.

Damper Fitted to slow movement in the suspension system, or act as a crankshaft balance (*see also* shock absorber).

Disc valve An induction system found on some two-stroke engines; usually a series of discs, located in the crankcase, rather than the cylinder barrel.

Displacement The engine capacity or volume displaced by the movement of the piston from its bottom-dead-centre position to top dead centre.

Deltabox A box-section, aluminium frame.

DOHC Double overhead camshaft. Two camshafts mounted in the cylinder head.

Dry sump An engine lubrication system, in which oil is contained in a separate reservoir, and carried to and from the engine by a pair of pumps.

Duplex frame A frame with twin main tubes to support the engine, rather than a single tube.

Earles fork A front fork design incorporating long leading links connected by a rigid pivot behind the front wheel.

Face cam A similar arrangement to an overhead camshaft, but without the cost and complexity.

Featherbed A Norton frame, designed by Rex McCandless.

Flat-twin An engine with two horizontally opposed cylinders.

Flywheel Fitted to the crankshaft, this heavy wheel smoothes intermittent firing pulses and helps smooth running.

Friction drive An early form of drive, using discs that press against each other instead of chains and gears.

Gearbox A case containing trains of gear wheels that can be moved to provide different ratios.

Gear ratio The ratio between the input and output speeds of a train of gears; indicates the difference between engine revolutions and rear wheel revolutions.

High camshaft A camshaft mounted high in the engine to shorten the pushrods in an overhead-valve design.

Hp Horsepower. A method of indicating engine displacement, used on early British machines; not to be confused with bhp.

Intercooling A system for cooling the fuel/air mixture as it passes from a turbocharger or supercharger to the cylinder, increasing power output.

IOE Inlet over exhaust. A common arrangement in early motorcycles: overhead inlet valve and side exhaust valve.

Leaf spring Several metal blades clamped together; used in early motorcycle suspension systems.

Liquid cooling A method of removing heat from engine cylinder barrels and heads by allowing water to flow through passages formed in them. Flow may be achieved by siphon or pump, the water passing through a radiator to cool it.

Magneto A high-tension dynamo that produces current for the ignition spark; superseded by coil ignition.

Main bearing A bearing in which the crankshaft runs.

Manifold A collection of pipes for supplying the fuel/air mixture to the engine, or for removing exhaust gases.

Monocoque Bodywork that doubles as a supporting structure for a machine's mechanical components.

OHC Overhead camshaft. A single camshaft mounted in the cylinder head.

OHV Overhead valve. An inlet or exhaust valve mounted in the cylinder head.

Piston Moves up and down the cylinder, drawing in the fuel/air mixture and compressing it, before being driven down by combustion to turn the crankshaft, then forcing out the exhaust gases; also used in hydraulic brake calipers.

Piston-port induction A method of supplying the fuel/air mixture to a two-stroke engine via the cylinder barrel.

Pushrod Operates an overhead valve by riding on a cam below the cylinder.

Reed valve Improves induction in a two-stroke engine via the cylinder; installed between cylinder and carburettor.

Shaft drive A final drive system, using a rigid shaft that runs between the engine and rear wheel.

Shock absorber Fitted to slow movement in the suspension system, or act as a crankshaft balance (*see also* damper).

SV Side valve. An inlet or exhaust valve mounted in the cylinder barrel.

Silencer Fitted to the exhaust system of an engine to reduce the pressure of the exhaust gases and lessen noise.

Swinging arm A form of rear suspension, in which an arm carries the wheel and is attached to the frame at its forward end by a pivot.

Torque Twisting force in a shaft; can be measured to determine at what speed an engine develops maximum pulling power.

Wet sump An engine lubrication system, in which the oil is contained within a sump that forms part of the crankcase.

INDEX

ORTONS
Media Group Ltd

Every so often a unique snapshot of times gone by is discovered in a dusty vault or in shoeboxes in an attic by an enthusiastic amateur photographer. They are living history. Each and every one of us cannot resist the temptation as we marvel at the quality of the images, to let our mind drift back to the good old days and wonder what it was really like.

We at Mortons Motorcycle Media, market-leading publishers of classic and vintage titles, own one of the largest photographic archives of its kind in the world. It is a treasure trove of millions of motorcycle and related images, many of which have never seen the light of day since they were filed away in the dark-room almost 100 years ago.

Perhaps the biggest gem of all is our collection of glass plates – almost two tons of them to be precise! They represent a largely hitherto unseen look into our motorcycling heritage from the turn of the century. Many of the plates are priceless and capture an era long gone when the pace of life was much slower and traffic jams were unheard of. We are delighted to be associated with well-known author Mick Walker in the production of this book and hope you enjoy the images from our archive.

Terry Clark,
Managing Director,
Mortons Media Group Ltd

"I am privileged to have access to the Mortons Motorcycle Media Archive, which I consider to be the finest in the world and a national treasure."
MICK WALKER

ACKNOWLEDGMENTS

The author would like to thank the following for their assistance:

Colin Seeley, for kindly agreeing to pen the Foreword
Alan Wilson of Redline Books
Gary Houlden of Webbs Yamaha Centre, Peterborough
Terry Rudd of TRM, Holbeach
Keith Davies of Three Cross Motorcycles
Jane Scayman of Mortons Media Group
Livio Lodi of the Ducati Museum, Bologna
The National Motorcycle Museum, Birmingham
Pooks Motor Bookshop, Leicester
Steve Bedford
Andrew Hunt
Philip Tooth
Martin Philp
Vic Bates
Doreen and John Hynes
Patty Ruston

PICTURE CREDITS